Handbook of National Communication Networks for Public Safety

The U.S. Department of Homeland Security created the FirstNet Authority for the purpose of building and operating a nationwide public safety broadband network. *Handbook of National Communication Networks for Public Safety: Next-Generation Advances* traces the evolution of the National Public Safety Broadband Network (NPSBN). The book is composed of three major sections, each section containg specific contents:

- Advent of FirstNet National Public Safety Broadband Network
- Evolution of National Public Safety Broadband Network (NPSBN)
- Future Innovations, Opportunities, and Challenges in Public Safety

The book covers planning, architecting/designing of networks and services, deployment, and operations, as well as enhancements to cover all states and territories of the United States. Other highlights covered in the book include:

- 4G/5G/6G mobile broadband technology
- The FirstNet and NPSBN ecosystem
- The dedicated and secure public safety network for first responders
- Terrestrial and satellite based broadband network and services for public safety
- 911 and location-based services
- Quality of service
- Cybersecurity
- AI/AR opportunities, network interoperability and challenges

The book features technical details of the NPSBN and services. It explains how enhanced devices make the NPSBN accessible anywhere and anytime to public safety personnel over dedicated and secure LTE/5G networks. It discusses how public safety challenges differ from urban to rural areas, from mainland to U.S. territories, and from land to sea coverage. The book concludes with an analysis on future challenges and presents approaches for successful resolution of these issues.

Madhu S. Singh, Ph.D., is an internationally recognized educator, innovator, and telecom executive having more than 40 years of experience in innovation, execution, and management of telecoms networks/products/services in wireless, wireline, and cable industries across globe. Since 2012, he has been instrumental in architecting, developing, and deploying the National Public Safety Broadband Networks (NPSBN) utilizing the dedicated spectrum B-14 to enable first responders to have access to a secure network anywhere anytime. He has also worked on expanding the NPSBN in rural areas across 50 states, Puerto Rico, the U.S. Virgin Islands, and the U.S. territories of Guam, American Samoa, and the Commonwealth of Northern Mariana Islands.

Handbook of National Communication Networks for Public Safety

Next-Generation Advances

Madhu S. Singh, Ph.D.

CRC Press
Taylor & Francis Group
AN AUERBACH BOOK

Designed cover image: Web Large Image (Public)

First edition published 2026
2385 NW Executive Center Drive, Suite 320, Boca Raton FL 33431

and by CRC Press
4 Park Square, Milton Park, Abingdon, Oxon, OX14 4RN

CRC Press is an imprint of Taylor & Francis Group, LLC

ISBN: 978-1-032-85899-9 (hbk)
ISBN: 978-1-032-90389-7 (pbk)
ISBN: 978-1-003-55773-9 (ebk)

DOI: 10.1201/9781003557739

Typeset in Minion
by Newgen Publishing UK

For their continued support and encouragement.
To my parents – Brahma Singh and Jagna Singh,
My wife – Kanti Singh,
My sons – Rajiv K. Singh and Sanjiv K. Singh,
My daughters-in-law – Dr. Anshu Singh and Dawn Tricarico,
And my loving grandchildren – Pooja K. Singh, Esq., Suraj P. Singh, Esq., and Mallika K. Singh.
Thank you for your unwavering support and encouragement.

Contents

Foreword

In the critical moments when disaster strikes and lives hang in the balance, public safety communication networks serve as the invisible lifelines connecting first responders, government agencies, and communities. These networks don't merely facilitate communication – they enable heroism, coordinate miracles, and transform chaos into coordinated response. Throughout his distinguished career, Dr. Madhu S. Singh has stood at the forefront of this mission-critical field, wielding innovation, resilience, and unwavering dedication to provide technical solutions in this area as his guiding principles.

From the ancient city of Varanasi, India, Madhu's extraordinary journey has carved a path across five continents, leaving an indelible mark on telecommunications and public safety infrastructure worldwide. His trajectory – from university lecture halls in India, Kenya, Australia, and the United States to commanding global leadership positions in telecommunications, public safety, and network security – exemplifies the relentless pursuit of excellence that defines true industry pioneers. Whether architecting FirstNet's groundbreaking National Public Safety Broadband Network (NPSBN) in the United States or orchestrating collaborative initiatives with industry titans across Asia, Africa, Australia, and Europe, his singular mission has never wavered: to architect communication networks that don't just connect voices, but preserve lives, revolutionize emergency response, and accelerate societal progress.

Across more than 40 transformative years, Dr. Singh has masterfully orchestrated the convergence of applied research, systems engineering, product development, and international standards development, fundamentally reshaping industry-wide best practices and setting new benchmarks for excellence. As an educator, mentor, and visionary industry contributor, he has cultivated generations of emerging professionals, authored seminal technical publications, and commanded respect within prestigious global standards committees, including the International Standards Organization (ISO), Telecommunication Industry Association (TIA), American National Standards Institute (ANSI), and National Institute of Standards and Technology (NIST). His comprehensive expertise encompasses compliance frameworks, sophisticated risk management protocols, cutting-edge broadband technology, and critical infrastructure protection – all indispensable pillars supporting next-generation public safety communications.

Transcending traditional corporate boundaries, Dr. Singh has personally engaged with stakeholders across all 50 U.S. states and territories, including Puerto Rico, the U.S. Virgin Islands, Guam, American Samoa, and Saipan, crafting bespoke solutions for extraordinarily diverse operational environments. His direct collaborations with 9-1-1 call centers, fire departments, emergency management teams, and satellite communications providers have forged an unparalleled understanding of operational realities, fundamentally strengthening his strategic approach to network architecture and seamless interoperability.

Today, through StarTech Global Consulting, he continues his relentless pursuit of excellence in 4G/5G networks and public safety solutions, ensuring that communication technologies remain not only secure and adaptive, but absolutely mission-critical for first responders operating under the most demanding conditions. Simultaneously, through his leadership as Director of Charity International, he champions the cause of underserved communities, masterfully bridging revolutionary technological advancements with profound humanitarian impact.

This handbook represents the crystallization of decades of global experience, unmatched technical insight, and strategic foresight – serving as an indispensable resource for engineers, policymakers, emergency responders, and researchers committed to designing, deploying, and optimizing national communication networks for public safety. As we confront increasingly sophisticated security threats, rapidly evolving technologies, and escalating demands for seamless interoperability, this comprehensive guide offers a robust framework for fortifying emergency communication infrastructures and empowering those who dedicate their lives to serving others. Through strategic knowledge-sharing, purposeful collaboration, and relentless innovation, we can architect networks that not only save lives and enhance preparedness, but also ensure that no call for help – regardless of location or circumstance – ever goes unanswered.

Dr. Satya Sharma, Ph.D.
Professor and Former Founding Executive Director,
Center of Excellence in Wireless and Information Technology,
Stony Brook University. Former Senior Vice President and General Manager,
Wireless and Mobile Computing and Head of Worldwide Operations,
Symbol Technology. Former Director of Bell Labs

Preface

Handbook of National Communication Networks for Public Safety – Next-Generation Advances

In an era where communication networks underpin every aspect of public safety, the need for robust, secure, and interoperable systems has never been more urgent. This handbook serves as a definitive guide to the next-generation advances shaping the future of national communication networks for public safety, providing insights into the transformative technologies that enable more responsive and resilient emergency services.

The genesis of this book stems from my extensive experience working at the intersection of telecommunications and public safety. From leading the FirstNet program's architectural developments to consulting on global telecom initiatives, I have witnessed firsthand the evolution of public safety communication networks. The rapid technological shifts – 5G, mission-critical push-to-talk, AI-driven network operations, and enhanced cybersecurity measures – signal both unprecedented opportunities and formidable challenges.

Extending public safety networks beyond the mainland to Puerto Rico, the U.S. Virgin Islands, Guam, Saipan, and American Samoa underscores the critical need for reliable communication infrastructure in island territories. These regions, often exposed to hurricanes, typhoons, and other natural disasters, demand resilient systems that can maintain emergency connectivity under extreme conditions. Strengthening telecommunications in these areas has been a deeply rewarding endeavor, allowing public safety agencies to improve response time and coordination in moments of crisis.

In scenarios where conventional 4G/5G towers are damaged, unavailable, or yet to be deployed, alternative technologies become lifelines for emergency response. Satellite communications, drone-based relay networks, and marine connectivity solutions play an indispensable role in enabling real-time situational awareness and ensuring first responders can reach those in dire need. Satellites provide a resilient backbone for emergency communication across vast territories, while drones equipped with network relay capabilities can restore connectivity in disaster zones. Also, marine communication systems enhance coordination for rescue operations, particularly in coastal regions where disasters strike the hardest.

At the heart of modern emergency response lies enhanced 9-1-1 (e9-1-1) services, which provide precise location tracking and multimedia communication capabilities – allowing first responders to receive images, videos, and real-time data from distressed callers. Further ensuring compliance and security, Communications Assistance for Law Enforcement Act (CALEA) services facilitate lawful interception and monitoring, ensuring that networks are equipped to support investigative and national security needs without compromising operational integrity.

Equally vital to public safety operations is the availability of dedicated communication handsets, purpose-built and certified for first responders. These ruggedized devices ensure uninterrupted connectivity under extreme conditions, supporting mission-critical voice, data, and multimedia communications essential for rescue and disaster response. With advancements in device interoperability and hardened security features, these specialized handsets stand as a cornerstone of next-generation emergency networks.

This handbook is designed to bridge the gap between technological advances and practical implementation, offering a comprehensive view of modern communication infrastructures tailored to the unique demands of emergency services. It provides policymakers, engineers, and public safety officials with the knowledge necessary to navigate the complexities of evolving network landscapes.

The chapters ahead delve into advancements in broadband communication for first responders, network resilience strategies, interoperability frameworks, and cybersecurity imperatives. I did explore not only the technological innovations but also the regulatory and operational considerations essential for ensuring that communication networks are reliable, secure, and adaptive in crisis situations.

It is my earnest hope and sincere desire that this book serves as a valuable resource for professionals dedicated to safeguarding communities through cutting-edge public safety communications. By fostering a deeper understanding of the challenges and breakthroughs in national communication networks, we move closer to a future where connectivity empowers swift, coordinated, and life-saving responses.

Acknowledgments

I would like to express my deepest gratitude to all those who have contributed to the creation of this book, *Handbook of National Communication Networks for Public Safety*. This work would not have been possible without the support, guidance, and encouragement of many individuals and organizations.

First and foremost, I extend my heartfelt thanks to my family for their unwavering support and understanding throughout the writing process. Their patience and encouragement have been invaluable.

I am profoundly grateful to my colleagues and mentors who have provided their expertise and insights, helping to shape the content and direction of this book. Their contributions have been instrumental in ensuring the accuracy and relevance of the information presented.

Special thanks to the various public safety agencies and communication network professionals who shared their experiences and knowledge. Their real-world insights have enriched the content and provided practical perspectives that are essential for the readers.

I also wish to acknowledge the support of my publisher whose belief in this project has been a driving force behind its completion. Their guidance and assistance have been crucial in bringing this book to fruition.

Finally, I would like to thank the readers of this book. Your interest and engagement are the ultimate reward for this endeavor. I hope that this handbook serves as a valuable resource in enhancing the understanding and implementation of national communication networks for public safety.

Thank you all for your contributions and support.

About the Author

Dr. Madhu S. Singh is a seasoned leader in public safety telecommunications, with a distinguished career spanning research, product and program management, and global consulting. As president of StarTech Global Consulting, he has spearheaded advancements in 4G/5G networks, ensuring robust emergency communication systems for first responders. His expertise in network security, compliance, and interoperability has played a pivotal role in shaping public safety broadband frameworks.

He held leadership positions at AT&T, Verizon, and Bellcore/Telcordia and brings a wealth of hands-on experience in developing critical communication infrastructure. His academic credentials include a Ph.D. and a master's degree in systems engineering, supplemented by a PMP certification and a mini-MBA. A passionate advocate for public safety innovation, he has contributed to global standards and policies aimed at strengthening emergency response networks. His work bridges technical precision with strategic vision, making complex network architectures accessible to policymakers and industry professionals. Beyond his technical expertise, Dr. Singh is committed to educating future leaders and fostering collaboration between public safety agencies and technology sectors. His *Handbook on the National communication Networks for Public Safety* offers a comprehensive exploration of mission-critical communications. Through this work, he aims to empower first responders, industry experts, and trainees with insights that enhance emergency response capabilities worldwide.

Use of the Book

This handbook serves as a comprehensive reference for professionals engaged in the innovation, design, implementation, and management of national communication networks for public safety. It provides a structured analysis of critical telecommunications frameworks, spectrum management policies, emergency broadcasting protocols, and evolving 5G/6G architectures relevant to public safety operations. Practitioners, policymakers, and researchers can leverage its insights to enhance interoperability, optimize network performance, and ensure resilient emergency communication systems. With detailed technical explanations, industry standards, and case studies, the book aids decision-makers in navigating regulatory requirements and advancing situational awareness in crisis response. The book also extends its relevance beyond the U.S. mainland, addressing communication challenges and solutions in territories such as Puerto Rico, U.S. Virgin Islands, American Samoa, Guam, and Saipan, ensuring robust public safety networks across all jurisdictions.

Additionally, this book serves as a valuable educational resource for students and trainers in public safety and emergency management, equipping them with foundational knowledge and real-world applications to foster expertise in the field.

Section I

Advent of FirstNet National Public Safety Broadband Network

1

Driving Force for an Advanced National Public Safety Network

Public safety refers to the welfare and protection of the public, often ensured by government organizations such as law enforcement, fire services, and emergency medical services. Its primary goal is to prevent emergencies, maintain peace, and protect citizens from danger or damage caused by natural or man-made events.

The driving force for a National Public Safety Network is the need for improved communication and coordination among emergency responders. Such networks aim to enhance the efficiency of public safety agencies, enabling them to share information seamlessly and respond more effectively to crises. This initiative is motivated by the increasing need for interoperability among different agencies and advancements in technology.

1.1 9/11 COMMISSION'S RECOMMENDATIONS FOR PUBLIC SAFETY

The 9/11 attacks in New York City and the August 2005 Hurricane Katrina in New Orleans highlighted the communication challenges faced by first responders during emergencies and disasters. The issues were emphasized in the 9/11 Commission Report, which identified gaps in emergency communications and recommended a nationwide network for law enforcement, fire, and emergency medical personnel communications. The public safety community has advocated for the implementation of the 9/11 Commission's recommendations to establish a dedicated and reliable wireless

DOI: 10.1201/9781003557739-2

network for first responders across all U.S. states and territories, including rural areas and tribal nations.

1.2 MIDDLE CLASS TAX RELIEF AND JOB CREATION ACT OF 2012 OF THE U.S. CONGRESS

On February 22, 2012, President Obama signed the Middle Class Tax Relief and Job Creation Act of 2012 (1).The Act established the First Responder Network Authority (also called simply FirstNet), an independent authority within the U.S. Department of Commerce's National Telecommunications and Information Administration (NTIA).

The Act authorized FirstNet Authority (FNA) to create Nationwide Public Safety Broadband Network (NPSBN) for first responders, covering all U.S. states, territories, rural areas, and tribal nations.

The U.S. Congress also initially allocated up to $7 billion dollars to FirstNet to construct NPSBN.

1.3 FEDERAL COMMUNICATIONS COMMISSION'S SPECTRUM ALLOCATION TO FNA

The Federal Communications Commission (FCC) is an independent agency of the U.S. government responsible for regulating interstate and international communications. Its roles include overseeing radio, television, wire, satellite, and cable communications, managing public airwaves, enforcing laws to promote fair competition, ensuring public safety communications, and facilitating broadband expansion and access.

The FCC reallocated spectrum and granted FirstNet license to both the 700 MHz D block (B14) and existing public safety broadband spectrum. The location of the 700 MHz band – just above the remaining TV broadcast channels – gives it excellent propagation characteristics. This allows the 700 MHz signals to penetrate buildings and walls easily and to cover larger geographic areas with less infrastructure (relative to frequencies in higher bands).

1.4 FIRSTNET BUDGET ALLOCATION

The U.S. Congress also initially allocated up to $7 billion to FirstNet to construct this NPSBN to provide a dedicated broadband network for public safety agencies in the United States and U.S. territories, including those in the Caribbean (Puerto Rico and U.S. Virgin Islands) and Pacific Region (Guam, Commonwealth of Northern Mariana Islands, and American Samoa). The provision includes authorization for additional funding to upgrade the network and services with innovative technologies, such as 5G networks, drones, and Low Earth Orbit (LEO) satellites.

1.5 DRIVING FORCE FOR AN ADVANCED NPSBN – KEY POINTS

The 9/11 terrorist attack in New York City and August 2005 Katrina hurricane Katrina in New Orleans, prompted the establishment of the 9/11 Commission. The commission identified a critical need of improved communication and coordination among emergency responders which prompted the U.S. Congress to enact the Middle Class Tax Relief and Job Creation Act of 2012. The Act established the First Responder Network Authority, also simply called "FirstNet."

The U.S. Congress initially allocated $7 billion to FirstNet to construct the NPSBN to provide a dedicated broadband network for public safety agencies in the United States and U.S. territories in the Caribbean (Puerto Rico and U.S. Virgin Islands) Pacific Region (Guam, Commonwealth of Northern Mariana Islands, and American Samoa). The provision includes on an ongoing basis additional fund allocation to upgrade the network and service incorporating new technology, e.g., 5G/6G network, drones, LEO satellite.

2

FirstNet Authority: Organizational Structure

FirstNet Authority (FNA) is a federal entity established to oversee the implementation and management of FirstNet. It aims to provide first responders with enhanced network connectivity and reliability, ensuring they can communicate effectively during emergencies. Organizationally, it is within the National Telecommunications and Information Administration (NTIA), under the U.S. Department of Commerce (DOC).

2.1 THE U.S. DEPARTMENT OF COMMERCE

DOC primary role is to promote economic growth, job creation, and U.S. competitiveness in the global market through trade promotion, economic analysis, and technological innovation. As part of the Executive Branch under the Office of the President, the DOC oversees various functions including domestic and international trade regulation.

The DOC fosters, incubates, and spearheads cutting-edge innovations and technologies to maintain United States' competitive advantage through its research and development entities and laboratories, including the National Institute of Standards & Technologies (NIST), NTIA, and the National Oceanic and Atmospheric Administration (NOAA).

DOI: 10.1201/9781003557739-3

2.2 NATIONAL TELECOMMUNICATIONS AND INFORMATION ADMINISTRATION

NTIA is the agency solely responsible for authorizing federal spectrum use and is the President's principal advisor on telecommunications policies, coordinating and communicating the views of the Executive Branch. The difference between NTIA and Federal Communication Commission (FCC) is that while NTIA manages federal spectrum for use by government agencies, the FCC is responsible for allocation and regulation of spectrum for nongovernment uses. The FCC is an independent federal agency, established through the Communications Act of 1934 by the U.S. Congress, and reports to Congress which provides its funding. Together, the FCC and NTIA jointly manage the nation's radio spectrum resources in the public interest.

NTIA is the parent agency of FNA and provides support and funding to FNA.

NTIA activities include:

- Managing the federal use of spectrum and identifying additional spectrum for commercial use.
- Administering grant programs that further the deployment and use of broadband and other technologies in the United States.
- Developing policy on issues related to the Internet economy, including online privacy, copyright protection, cybersecurity, and the global free flow of information online.
- Promoting the stability and security of the Internet's domain name system through its participation on behalf of the U.S. government in Internet Corporation for Assigned Names and Numbers (ICANN) activities.
- Engaging in advanced telecommunications research and engineering in collaboration with the federal government and private sector partners.

2.3 FNA BOARD

The Middle Class Tax Relief and Job Creation Act of 2012 established a governing board for FirstNet. This board authorizes FirstNet's management team and program office to operate the organization. The board also provides short and long-term strategic direction, makes enhancements and network reinvestment decisions, approves the annual budget, and authorizes all major contracts.

The board is composed of individuals with expertise in management, finance, communication networks, and public safety. They are dedicated to making the NPSBN successful. Currently, the board performs its work thorough various committees, namely Advocacy, Finance and Investments, Governance and Risk, and Program and Future Planning. The members may serve on more than one committee.

The FNA Board currently has nearly 15 members, including representatives from the Secretary of Homeland Security, the Attorney General, and the Office of Management and Budget. The remaining members are chosen by the Secretary of Commerce based on their expertise in public safety, technical fields, networks, and/or finance.

2.3.1 FNA Oversight Committee

U.S. Congress's subcommittee on Communications, Technology, Innovation, and the Internet have periodically convened to know the stages of works related to NPSBN (2). The FirstNet Oversight Committee is responsible for overseeing the implementation and operations of FirstNet. This committee ensures that the network is efficient in public safety communications and provides guidance on governance and strategy.

2.4 FNA – PSAC

In addition to the Board, FNA also receives guidance from the Public Safety Advisory Committee (PSAC) (3) as described here:

- Provides subject matter expertise on concepts (e.g., policies, procedures, technologies, operational methods) developed by FirstNet.
- Advises FirstNet (FN) through the creation of initial documents, plans, or reports related to the build-out, deployment, and operation of NPSBN. The PSAC reports include subject matter related to intergovernmental responsibilities or administration which are shared to enhance public safety.

- Conducts regular meetings to remain informed and up to date on FN progress.

2.5 FNA'S COORDINATION WITH NIST

FN closely coordinates the strategic technical development work with the Public Safety Communication Research Division of the NIST.

2.5.1 National Institute of Standards and Technology

NIST is an agency within the U.S. DOC. Its mission is to promote innovation and industrial competitiveness by advancing measurement in science, standards, and technology. NIST works and coordinates with industries, universities, government and private laboratories in developing new technical knowledge including Generative AI, improving existing technologies, and creating standards.

In addition, NIST develops standards and guidelines that assist sectors such as information technology, cybersecurity, manufacturing, and healthcare. It supports industries through several programs by providing laboratory facilities, technical assessments, and program evaluations.

2.5.2 FN Boulder Lab

FN lab, located in Boulder, Colorado, is owned and operated by FN in coordination with NIST aligning the Telecommunications Industry Association (TIA) standards. The FN lab is fully equipped to evaluate public safety functionalities and features unique to the FN network including Quality of Service (QoS), priority and preemption, enhanced situational awareness technologies and applications. The lab also has full capability to test and evaluate mobile devices, new network elements, futuristic services, and applications.

The FNA and NIST plan to update the facility regularly with enhancements such as haptics and metrology, 5G/6G, edge computing, XR capabilities – Extended Reality – which allow virtual and augmented realities to be combined) and other technological advances for public safety operations. Through this partnership, NIST's Public Safety Communications Research (PSCR) division and the FNA plan to enable research and development, education, and training, by offering the facility at no cost to public safety agencies and organizations that support public safety response efforts, including private sectors and academic institutions. The facility is designed to help answer key research questions concerning the future of user interfaces and location services for public safety training and operations.

2.6 FNA ORGANIZATIONAL STRUCTURE – KEY POINTS

FNA is a federal agency responsible for the NPSBN in the United States and its territories. The Middle Class Tax Relief and Job Creation Act of 2012 mandated to establish FNA within the U.S. DOC under the NTIA which acts as the U.S. President's principal advisor on telecommunications policies, coordinating and communicating the views of the Executive Branch.

FNA works and coordinates its plans and programs with federal, state, local, and tribal public safety organizations. FNA is guided by its FN Board, the FN Oversight Committee, and the FN Advisory Committee.

FN closely coordinates with the PSCR division of the NIST.

3

Public and Private Handshake for Public Safety NPSBN

The advent of the National Public Safety Broadband Network (NPSBN) is truly a unique public and private "handshake" endeavor. It is a collaborative agreement or partnership where two entities work together on a specific project or initiative. This can involve joint partnerships or strategic alliances that leverage the strengths of both types of organizations. For example, a private company may provide innovative technology while a public organization may offer resources and wider market access.

The unique handshake endeavor built a highly reliable, dedicated, efficient, and secure wireless broadband network to cater to the public safety's immense needs to manage the natural or man-made disaster efficiently and save the lives and properties. From the lessons learned from the terrorist attack on World Trade Center in New York City on September 11, 2001, Hurricane Caterina, and many such disasters, it was clear that the NPSBN task would require a close coordination among government, public safety agencies, private sector network service, and equipment providers.

3.1 PUBLIC SAFETY AGENCIES

In the United States and many other countries, public safety is not overseen by a single agency. They are managed by local, county, state, or federal bodies that offer firefighting, law enforcement, ambulance, medical, and other emergency services.

Public safety agencies typically own and manage their local communication networks, devices, and protocols, but have none to very little

DOI: 10.1201/9781003557739-4

intercommunication networks among them to coordinate emergencies. The FirstNet Authority's (FNA) NPSBN provides a single communications network platform to public safety agencies to coordinate and manage emergency responses effectively by sharing the data, video, locations and other relevant information in real time.

3.2 ROLE OF AMERICAN TELEPHONE AND TELEGRAPH COMPANY (AT&T) IN NPSBN

The FNA awarded a contract to American Telephone and Telegraph Company (AT&T) to build NPSBN after a competitive Request for Information (RFI) and Request for Proposal (RFP) submitted by many well-known telecoms and information technology companies. Thus, AT&T became part of the "public and private partnership" to develop and expand the NPSBN and to reaching out to public service agencies and supporting them with broadband network and services in the United Stats as well as its territories.

AT&T is an international telecommunication company with a long history of being a premier U.S. telecommunications service provider. Alexander Graham Bell, the telephone inventor, got a patent by the U.S. Patent Office in 1876 and demonstrated his ability to transmit a call on a wireline. Graham Bell established Bell Telephone Company which later evolved into American Telephone and Telegraph Company.

The congressional mandate to FNA includes 99% of the U.S. population and 2.9 million square miles of coverage by NPSBN. FNA's selection of AT&T as a partner to build the NPSBN was definitely a smart decision since AT&T along with its subsidiaries and partners, has large coverage areas in the United States and its territories and serves both public and private enterprises as well as the general public.

3.3 RURAL COMMUNICATIONS SERVICE PROVIDERS

Rural communications service providers are regional telecom companies serving rural areas where AT&T lacks wireless coverage. In the United States, the major rural service providers are Cover, Panhandle, Bug Tussel,

TABLE 3.1
FNA/AT&T Rural Service Partners

State	FirstNet Carrier Owner
IA	Upper Mid-West
	Chat
	Cover 2
IL	Cover 2
MN	Cover 2
OK	Panhandle
TX	Panhandle
WI	Bug Tussel-A
	Bug Tussel-C

and Upper Mid-West and serve the rural areas of Iowa, Illinois, Minnesota, Oklahoma, Texas.

Generally, these service providers have wireline and wireless networks and serve the rural populations by providing basic phone and wireless services but have no network capacity and technologies to provide the public safety services mandated by FNA. AT&T has partnered with such rural communications providers, provides them use of FirstNet B14 spectrum, network facilities and operations support to service first responders. Table 3.1 provides a sample list of rural partners and their service states.

3.3.1 Public Safety Agencies and NPSBN Partnership in the U.S. Territories

The Middle-Class Tax Relief and Job Creation Act of 2012 has also provision to support first responders in the U.S. territories. Within the FNA public and private handshake framework, AT&T has contracts with the wireless service providers in these territories to support first responders and provide FN services in their respective service areas (see Table 3.2).

3.3.2 Radio Access Network, Small Cell, and Satellite Networks for Public Safety

During an emergency situation first responders may encounter areas where network coverage is temporarily unavailable or there is no network coverage available at all. Small cell and satellite networks provide network access to first responders.

TABLE 3.2
FirstNet Services in U.S. Territories

Geolocation	Territories	Service Providers	Remarks
Caribbean Island	Puerto Rico	Liberty Latin America	Liberty Latin America initially bought AT&T wireless networks in Puerto Rico and U.S. Virgin Islands
Caribbean Island	U.S. Virgin Islands, St. Croix, St. John, and St. Thomas	Liberty Latin America	
Pacific Island	Guam	Nippon Telegraph and Telephone Corporation's Do Communications Over the Mobile Network (DOCOMO)	DOCOMO is a subsidiary of Nippon Telegraph and Telephone Corporation of Japan
Pacific Island	Commonwealth of Northern Marian Islands	DOCOMO	
Pacific Island	American Samoa	American Samoa Telecommunications Authority	

Small Cell

Small cell, a wireless access network node, is generally deployed to provide coverage inside buildings and localized areas where the wireless signal from the Long-Term Evolution (LTE) or 5G Radio Access Network could not be reached or the signal is very weak. Generally, the small cell coverage is within 1,000–1,500 feet (300–450 meters) depending upon factors such as manufacturer and environment. First responders use their wireless devices to connect through small cells in such situations and provide the desired services.

Satellite Communication Networks Providers

Satellite communication networks use satellites to transmit signals for various forms of data communication, including television, internet, and radio. Satellites receive signals from ground stations, amplify them, and then relay them back to other ground stations or directly to users. This technology enables long-distance communication over vast geographical areas, making it essential for global telecommunications, broadcasting and navigation systems.

Satellite networks supplement FirstNet NPSBN coverage in remote areas where terrestrial radio access is unavailable or disrupted (e.g., wilderness search and rescue, firefighting, law enforcement activities in rural areas). Satellite services also maintain communication during terrestrial network restoration. With the recent advent of Low Earth Orbit (OEB) satellites, both Geostationary Earth Orbit (GEO) and LEO satellites are used for communications.

3.4 PUBLIC AND PRIVATE HANDSHAKE FOR NPSBN – KEY POINTS

In the United States, the responsibility of public safety is not under a single governing agency. Basically, the public safety agencies are functionally managed by local municipality, county, the state, or the federal agencies. These public safety agencies generally own and manage their own local communications networks, devices, and operational protocols, and have none to very little intercommunication networks among them to coordinate emergencies. The advent of the NPSBN is truly a unique public and private "handshake" endeavor where the FNA partners with AT&T to develop and manage NPSBN.

The FNA and AT&T partnership to build the NPSBN was a critical step in the right direction. This handshake enabled the FNA to serve public safety agencies nationwide, enabling public safety providers to enhance existing communication infrastructure within a relatively brief timeframe to meet critical requirements for saving lives and protecting properties.

Section II

Evolution of National Public Safety Broadband Network (NPSBN)

4

Charter, Strategies, and Scope of NPSBN

Public safety requires a nationwide interoperable broadband network that covers urban, suburban, and rural areas and meets the information and communications technology needs associated with public safety's mission. In addition, FirstNet, through its private business partner, American Telephone and Telegraph Company (AT&T), will continue to upgrade and innovate the National Public Safety Broadband Network (NPSBN) as Long-Term Evolution (LTE) and wireless broadband technologies (e.g., 5G, 6G) evolve, as public safety needs expand, and as new capabilities and technologies become accepted and available.

The FirstNet objectives for NPSBN services include (4):

1. Building, development, operation, and maintenance of the NPSBN: Provide NPSBN that ensures network coverage 24 hours a day, 7 days a week, 365 days a year and complies with the 3GPP and other national and international standards technical specifications.
2. Financial sustainability: Maximize the impact of government funding and leverage all 20 MHz of band 14 to build, deploy, operate, and maintain the NPSBN to serve public safety and for secondary use while ensuring a self-sustaining business model.
3. First responder user adoption: Establish (i) compelling, differentiated, and competitively priced service packages and (ii) sales, distribution, and marketing capabilities to ensure adoption of FirstNet products and services by a majority of eligible Public Safety Entities (PSEs).
4. Device ecosystem: Provide and maintain a 3GPP-compliant, band 14-capable device portfolio that evolves with the 3GPP standards and provides functionality and price points that meet the needs

DOI: 10.1201/9781003557739-6

of the FirstNet public safety customer base and drive substantial subscribership. FirstNet anticipates NPSBN customers will expect both mass market and ruggedized devices that are capable of gloved, one-handed, or hands-free operation as well as multimedia and high-definition data transmission from both humans and machine-based sensors.

The ecosystem must support Bring Your Own Device (BYOD) as well as, at a minimum, devices that:

- Operate seamlessly on the NPSBN and roam onto networks, including non-band 14 commercially available networks.
- Interoperate with FirstNet's applications ecosystem.
- Support Universal Integrated Circuit Card (UICC) features and options, including the ability to home and, if applicable, roam to multiple networks while prioritizing them appropriately.
- Operate seamlessly with a comprehensive device management system to allow remote provisioning and control.

5. Application ecosystem: Provide an applications ecosystem that supports the NPSBN with capabilities and services relevant to public safety. The ecosystem includes, at a minimum:
   - An evolving portfolio of mobile and enterprise applications, as well as cloud services.
   - An applications development platform.
   - A vibrant third-party applications developer community.
   - An applications store.
   - Local control of users, subscriptions, services, and applications.
   - User-friendly federation of identity management.
   - Data, application, and resource sharing across diverse PSEs.
   - Core service and application delivery platforms.
   - Data and applications security and privacy compliance across local, tribal, state, regional, and federal users.
6. User service availability: Provide a broadband service with 99.99% availability, exclusive of planned maintenance windows, as measured in a rolling 12-month window within each reporting area. The NPSBN partner must consider areas that contain mission-critical infrastructure as needing enhanced hardening and increased availability. Service restoration activities undertaken with the highest available priority but without exceeding two hours for any impaired service. For restoration of service via temporary or secondary service capabilities, the

temporary or secondary service must be transparent to the users and provide similar capability.

7. Service capacity: Provide service capacity to support geographically dispersed public safety usage including a map illustrating first responder density and current mobile data usage.
8. Cybersecurity: Provide cybersecurity solutions using the extensive set of industry standards and best practices contained in the International Organization for Standardization (ISO) and International Electrotechnical Commission (IEC) security guidance for networks.
9. Providing encryption capabilities to support federal, state, and local public safety users. Protective measures need to be applied end to end across the FirstNet enterprise environment and will include securing user equipment (UE), applications running on UE, the Radio Access Network (RAN), and the core network. In addition, considerations to support Identity, Credential, and Access Management in a user-friendly, secure, and federated fashion will be key to any successful cybersecurity solution.
10. Priority services: Provide a solution that allows the assignment of Quality of Service (QoS), priority, and preemption (QPP) parameters to user profiles through standard service control parameters defined by 3GPP and the Internet Engineering Task Force, including Access Class, Quality Class Indicator (QCI), Allocation and Retention Priority (ARP), and Differentiated Service (DiffServ). Allow control and management of static and dynamically assigned QPP parameters for public safety users and the ability to change user profiles in real time in response to incidents. User profile assignments and changes should be managed locally by PSEs.
11. Life-Cycle innovation: Evolve the NPSBN solution, including products and services, and incorporate 3GPP LTE/5G standards as they evolve and mature through the contract.

4.1 NPSBN COVERAGE

America's public safety network surpasses 2.97 million square miles, expands first responder access to 5G and strengthens commitment to mission-ready infrastructure in the U.S. Mainland. FirstNet wireless services are also

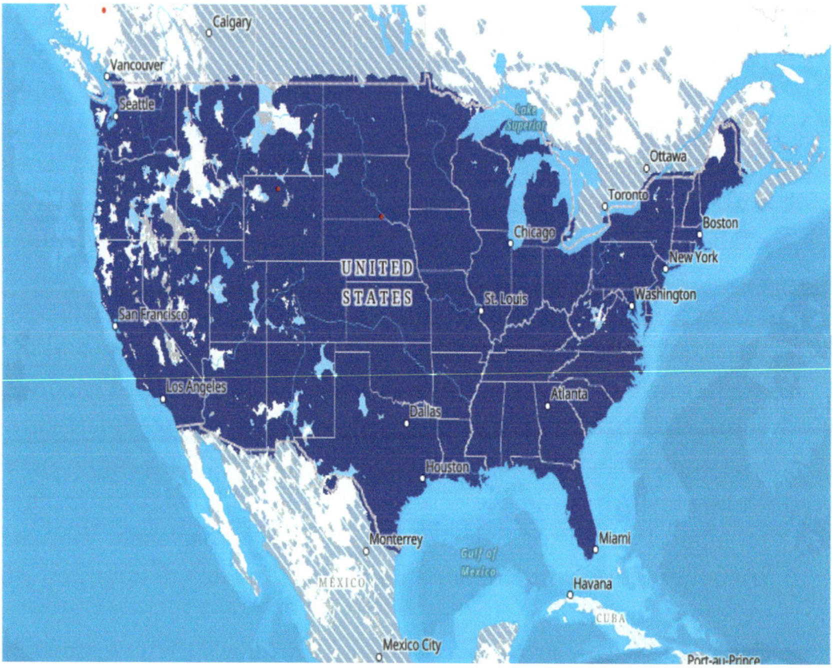

FIGURE 4.1
Current FirstNet NPSBN Coverage in the U.S. Mainland.

available in Alaska, Hawaii, and the U.S. terriories in the Caribbean and Pacific regions.

Wireless coverage refers to the geographic area where a base station and mobile device can reliably communicate with each other above a minimum designed data rate. Persistent and temporary coverage is defined as an LTE band 14 network capable of providing cell-edge data rates of 256 kbps uplink (UL) and 768 kbps downlink (DL) measured from outdoor stationary UE at three feet from ground level with a 95% confidence margin for the cell area with a uniform cell load of 50% for the DL and UL. The link budget cell-edge data rates (256 kbps UL/768 kbps DL) are primarily minimum design targets used to ensure overlap between cells is sufficient to maintain the minimum grade of service. Most user speeds will be significantly higher as user devices are likely distributed throughout a cell and speeds generally increase closer to the cell site. Figure 4.1 illustrates the NPSBN coverage areas depicted in grey color. The white areas are typically covered by FirstNet/AT&T's rural partners.

4.2 NPSBN SERVICES

The FirstNet charter required that the NPSBN support all existing voice, data, and video services available to public safety and first responders over the LTE network. In addition, the FNA will work with AT&T to introduce new services. Following is a list of typical services currently supported by NPSBN:

- Voice over LTE (VoLTE) encompasses support for all current voice call categories and functionalities, such as call forwarding, call waiting, and three-way calling, among others.
- Data services including texting.
- Video conferencing including Mission Critical Push to Talk (MC-PTT).
- Emergency calls (9-1-1, e9-1-1).
- Real-Time Texting (RTT) calls enabling communication in both text and voice replacing Teletypewriters/Telecmmunications Device for the Deaf (TTY/TDD).
- Wi-Fi Calling.
- Wireless Emergency Alert (WEA).
- Wireless Priority Service (WPS).
- Multimedia Services Synching between mobile handheld and mobile wearable devices.

FNA continuously strives to bring futuristic services to enhance the first responder's ability in their day-to-day activities.

4.3 NPSBN DEVICES

One of the basic requirements for all wireless devices connecting to NPSBN to receive desired service must be FirstNet enabled and certified having a FirstNet SIM. Any device which is not FirstNet enabled could not access the secure and priority-based services over NPSBN. FirstNet works hand in hand with AT&T and device vendors to certify the wireless devices, enhance the devices as needed, and develop new devices. AT&T maintains a catalog and inventory of all FirstNet-certified devices accessible to PSEs. It should be noted that first responders may acquire an FirstNet-certified wireless device through their affiliated public service agencies or from AT&T.

To ensure that these diverse group of users have the capabilities to perform their day-to-day activities, FirstNet supports a wide variety of approved devices, including Android®, and Apple® iOS smartphones, feature phones, and data-only devices, such as tablets, laptops, air cards, netbooks, and mobile hotspot devices.

4.4 NPSBN USER COMMUNITIES

FirstNet was initially designed exclusively for the "primary" use's category of PSEs such as police, fire, ambulance, and Emergency Medical Technicians (EMTs). However, many other services – as well as critical infrastructure – are often needed during emergencies and in the aftermath of a catastrophic event. For that reason, FirstNet was extended to include additional users. FirstNet user communities are listed below:

A: Primary Users Category includes:
- Police
- Fire
- Ambulance
- EMT

B: Extended Primary Users Category includes both public and private organizations:
- Electric utilities and power grid infrastructure
- Telecommunication services
- Hospitals and medical facilities
- Schools
- Water supply and wastewater treatment plants
- Transportation systems including airports, commercial rail systems, and public transit

C: Secondary Users Categories includes those who are not involved generally in public safety activities; however, at the time of the emergency such users may be included in emergency control management team and provided FirstNet-enabled devices for the specific event alone.

4.5 NPSBN PRIVACY AND SECURITY

NPSBN emphasizes privacy and security through its design and operational protocols. It is designed to protect sensitive data, secure communications, and provide robust access controls to ensure that only authorized personnel can access and use the network. The infrastructure is built to withstand potential cyber threats, focusing on encryption, user authentication, and compliance with federal guidelines to safeguard public safety communications.

4.6 INTEROPERABILITY AND OPERATIONAL INNOVATIONS

Interoperability refers to the ability of different systems and organizations to work together seamlessly, allowing for effective communication and data exchange. Operational innovations involve new methods and practices that improve efficiency, productivity, and outcomes within organizations. When combined, interoperability and operational innovations can enhance collaboration, streamline processes, and lead to more effective service delivery across various sectors.

4.7 CHARTER, STRATEGIES, AND SCOPE OF NPSBN – KEY POINTS

The Middle Class Tax Relief and Job Creation Act of 2012 mandated the NPSBN meet the information and communications technology needs associated with public safety's mission.

The scope of the NPSBN includes 4G/5G technology-based network, broadband-based services, and coverage in urban, suburban, rural, and tribal regions in the United States and its territories in the Caribbean and Pacific regions. Also, the infrastructure provides safeguard to potential cyber threats, focusing on encryption, user authentication, and compliance with federal guidelines to safeguard public safety communications.

FirstNet serves not only the "primary" category of PSEs such as police, fire, ambulance, and EMTs, but also provides access to NPSBN for managing critical infrastructures and supports during emergencies and in the aftermath of catastrophic events.

5

The FNA National Public Safety Broadband Network

FirstNet Authority's (FNA) National Public Safety Broadband Network (NPSBN) is a wireless broadband communications network. A wireless network is composed of user devices (e.g., mobile phones, tablets, laptops), Radio Access Network (RAN), core network, transport network, Operational Support Systems (OSS), and Business Support Systems. Details of each component are discussed in the later sections of the book.

Modern wireless telecommunications network complements towards flexible and efficient packet switching technologies compared with traditional circuit-switched networks worldwide using radio spectrum to connect and transmit signals between users (the caller and the receiver of the call). The caller using a wireless device connects to a radio tower. The signal processing is performed within the network components to establish the communications and exchange of multimedia information between the callers and the receivers. Communication may be established between person to person, machine to machine, or any other combination.

5.1 NPSBN LOGICAL REFERENCE ARCHITECTURE

As discussed in Chapter 4, the U.S. Congress mandated building NPSBN to support current and future public safety-related services, to ensure the economic sustainability of the NPSBN, and to enhance interoperability among enterprises, both public and private. The NPSBN is based on the 3GPP Long-Term Evolution (LTE) standards and will continue progressing using national and international standards. LTE is a standard for wireless broadband

DOI: 10.1201/9781003557739-7

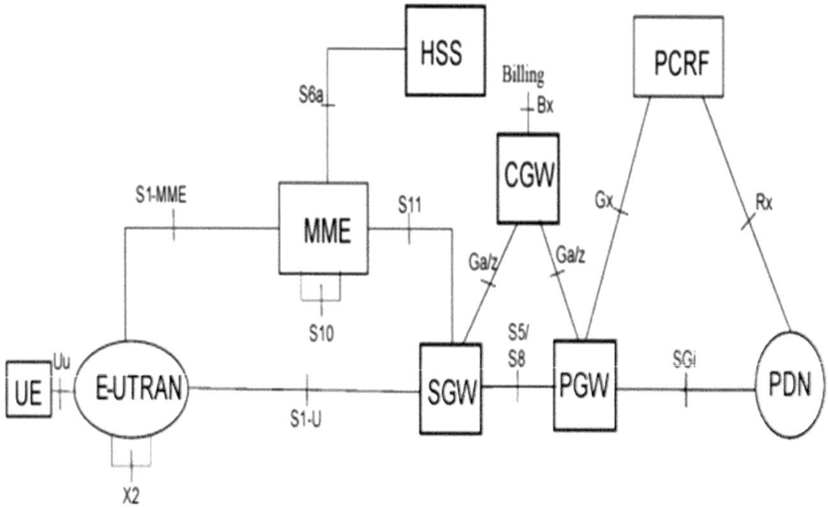

FIGURE 5.1
3GPP LTE Logical Reference Architecture.

communication that is often referred to as 4G. In this book "LTE" and "4G" are used synonymously.

A high-level architecture of 3GPP Logical Reference Architecture is shown in Figure 5.1 (5). It depicts the interconnections among various 4G wireless communications components and their interface protocols. The LTE networks mainly consist of the following major components and subsystems.

5.1.1 User Equipment

The user equipment (UE), which is also generally called mobile devices, attaches over air to an LTE RAN to send and receive voice, data, and video. The LTE UE has the following core modules:

- All communication functions are managed by Mobile Termination (MT).
- The data streams are terminated in the Terminal Equipment (TE)
- The UE equipped with a SIM card for LTE devices uses what is known as the Universal Subscriber Identity Module (SIM) and Universal Integrated Circuit Card (UICC). This application is known as the Universal Subscriber Identity Module (USIM), and the information stored on a USIM card includes user's telephone number, home network identity, and security key.

5.1.2 Evolved Universal Terrestrial Radio Access Network

Evolved Universal Terrestrial Radio Access Network (E-UTRAN) is also known as LTE/4G RAN. It consists of evolved base stations called eNodeB (eNB) which manages and controls radio signals and communicates with UE, Evolved Packet Core (EPC), and many subsystems to provide seamless connectivity and efficient data transmission in the LTE networks.

5.1.3 Evolved Packet Core (EPC)

The EPC is the heart of the LTE/4G network architecture. EPC communicates with external Packet Data Networks (PDN) such as the internet, private corporate networks, the IP multimedia subsystem, and Private Service Telephone Network (PSTN). The EPC is responsible for authentication, attachments, data transmission control and management, Quality of Service (QoS), and the network security.

Figure 5.1 illustrates the UE and E-UTRAN connectivity with EPC's nodes over a specific interface. Mobility Management Entity (MME) node is primary for managing attachment, authentication, security, and interfacing with Home Subscribers Service (HSS). Service Gateway (SGW) and Packet Gateway (PGW) nodes are primarily responsible for the packet data communication between PDN and the UE based on the subscribers' service quality profiles stored in Policy and Charging Rule Function (PDRF) node. Computing Gateway (CGW) node is interface with BSS to generate appropriate billing. The functional details of these nodes and their interfaces are discussed latter in LTE core section 7.1.

5.1.4 Federal Communications Commission (FCC) Radio Spectrum

Radio spectrum is electromagnetic wave having a wide frequency range from 1 Hz to 3,000 GHz (3-THz). Information is processed and carried over the designated electromagnetic wave range depending upon the Federal Communications Commission (FCC) spectrum allocation. Radio spectrum is one of the most valuable assets which generate revenue for the government.

Radio spectrum is used for numerous purposes including mobile phones, television broadcasts, cable television broadcasting, microwave oven, radar systems, microwave devices/communications, radio astronomy, wireless

LAN, Bluetooth, Zigbee, GPS, and two-way radios such as land mobile, emergency locator beacons, amateur radio, and satellite radio.

In the United States, the Federal Communications Commission (FCC) is an independent federal regulatory agency reporting directly to Congress. The FCC is the sole authority to allocate radio spectrum to various uses. The FCC is responsible for implementing and enforcing America's communications law and regulations. It also regulates interstate and international communications by radio, television, wire, satellite, and cable in all 50 states, the District of Columbia, and the U.S. territories.

5.1.5 FNA's Licensed Spectrum – Band 14

The 700 MHz band is an important swathe of spectrum available for both commercial wireless and public safety communications. Spectrum behaves differently – lower frequency travel farther but have more background noise, referred as "white noise" or "floor noise." Higher frequencies travel shorter distances but have better sound quality and less background noise.

Sections of spectrum are called "bands." This band consists of 108 MHz of spectrum running from 698 to 806 MHz and was freed up as a result of the Digital Television Transition. The location of the 700 MHz band – just above the remaining TV broadcast channels – gives it excellent propagation characteristics. This allows the 700 MHz signals to penetrate buildings and walls easily and to cover larger geographic areas with less infrastructure (relative to frequencies in higher bands).

In 2008, the FCC auctioned licenses to use portions of the 700 MHz band for commercial purposes. Mobile wireless service providers have since begun using this spectrum to offer mobile broadband services for smartphones, tablets, laptop computers, and other mobile devices.

The Middle Class Tax Relief and Job Creation Act of 2012 (Spectrum Act) directed the allocation of D Block (758–763 MHz/788–793 MHz), also called Band 14, to public safety for use in a nationwide broadband network. FirstNet holds the license for both the existing public safety broadband spectrum (763–769 MHz/793–799 MHz) and the reallocated D Block.

Figure 5.2 shows the spectrum allocated in the 700 MHz band after enactment of the Spectrum Act. It depicts two 10 MHz between 758–768 MHz and 788–798 MHz, in total 20 MHz spectrum allocation to FirstNet for NPSBN.

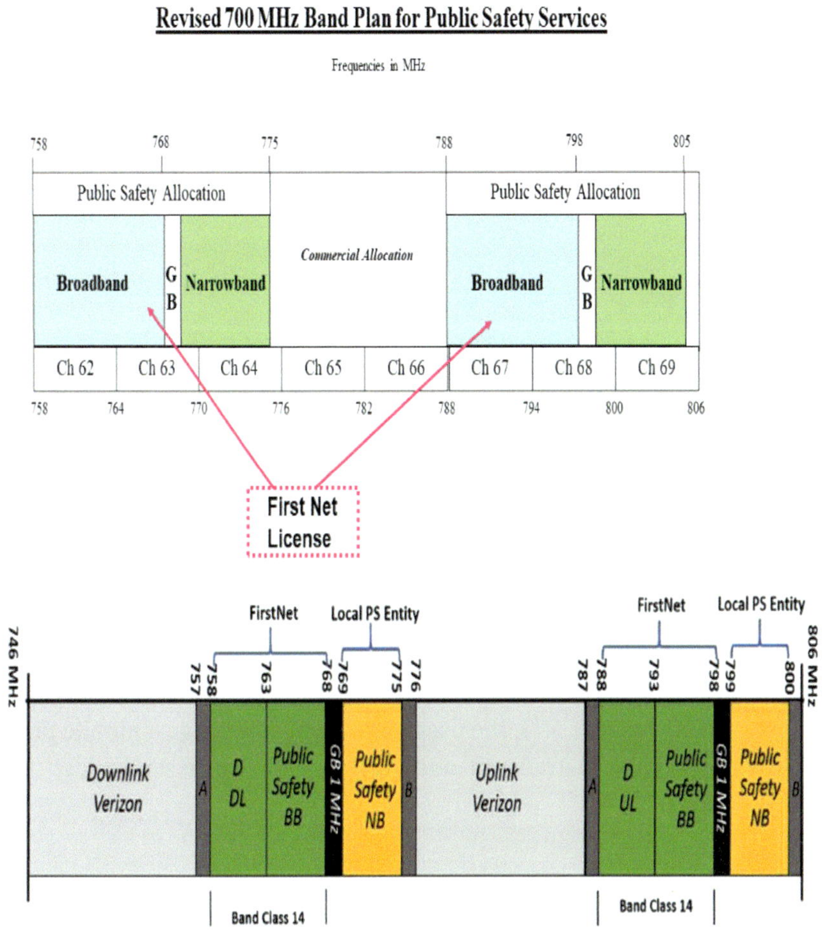

FIGURE 5.2
FirstNet NPSBN Spectrum Allocation.

5.2 NPSBN – KEY POINTS

FNA's NPSBN is based on 4G/5G-based wireless technology and conforms to 3GPP and many national and international standards. NPSBN uses 700 MHz band radio spectrum specifically allotted to FNA by FCC for public safety uses.

6

NPSBN Radio Access Network (RAN)

The Radio Access Network (RAN) is a part of National Public Safety Broadband Network (NPSBN) that connects FirstNet (FN) devices to the FN core network and facilitates wireless communications. It consists of radio antenna, Remote Radio Head (RRH), and base stations and their interconnections, allowing mobile users to access services such as voice, data, and video transmission.

As illustrated in 3GPP Long-Term Evolution (LTE) Logical Reference Architecture (see Figure 5.1), the user equipment (UE), e.g., a mobile phone, connects to a RAN to transmit and receive data, voice, and video from NPSBN core.

6.1 NPSBN LTE/4G RAN

NPSBN LTE/4G RAN consists of three major components: antenna, RRH, and Baseband Unit (BBU). Figure 6.1 illustrates a typical NPSBN RAN tower hosting the antenna and RRH. The BBU is always typically located at the base of the tower. Antenna and RRH connect with BBU using fiber optical cables.

- Antenna: Transmits/receives Radio Frequency (RF) signals over the air interfaces with mobile devices wirelessly. In NPSBN RAN, the omni-directional antenna transmits/receives RF from all directions and has Multiple Input Multiple Output (MIMO) features enabling attaching a large number of UEs at the same time. The antenna impacts the coverage areas of cell tower depending upon power, design, and configuration.

DOI: 10.1201/9781003557739-8

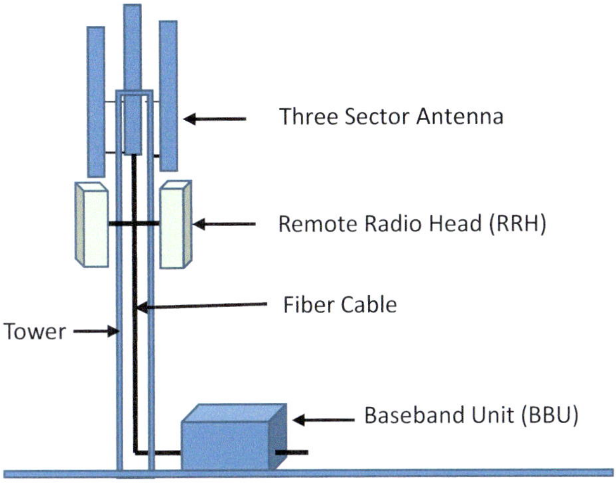

FIGURE 6.1
NPSBN LTE/4G Typical Tower Installation.

- RRH: Also called Remote Radio Unit, connects with antenna on one side and with the BBU on the other using a fiber optical cable. RRH converts the RF signal into data signal and vice versa. It also performs the filtering and amplification of RF signal. The RRH is usually mounted near the antenna to reduce transmission line losses.
- BBU: Connects the LTE EPC core over cable and manages the whole base station including operational maintenance as well as signal processing.

Figure 6.2 depicts a logical RAN architecture which consists of eNBs and interfaces connecting to LTE core elements, namely Mobility Management Equipment (MME) and Serving Gateway (SGW).

- An eNB supports user plane (S1C) and control plane (S1U) interfaces connecting to MME and SGW, respectively. The control plane interface performs mobility session management, authentications, attachments, signal bearer management, roaming, and handover functions. The use interface (S1U) performs user data exchange between SGW and the eNB to continue transmission to user devices (UE). Each interface uses a specific interface protocol. An interface defines how two entities/nodes may communicate. A protocol defines how they should communicate and what that communication means. A wireless UE, e.g., mobile phone, communicates to eNBs over air using UE interface. In addition, eNBs

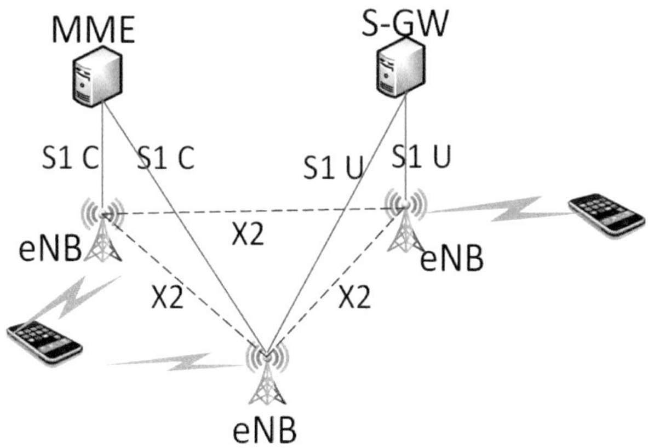

FIGURE 6.2
4G/eNB Interface Logical Architecture.

are interconnected with their adjoining eNBs over the X2 interface. X2 supports enhanced mobility, intercell interface management, and SON functionalities; eNBs are connected through S1 interface to the Evolved Packet Core (EPC) (6).

The eNB communicates with key LTE/4G network subsystems and performs the following tasks:

Radio Access Point (RAP)

The primary function of the eNB is to serve as a radio access point that connects UEs to the LTE network. It communicates with UEs over the air interface, managing the radio resources and ensuring that wireless connections are established and maintained.

Air Interface Protocol Stack

The eNB is responsible for implementing the lower layers of the LTE protocol stack, including the Physical Layer (Layer 1) and the Medium Access Control (MAC) layer (Layer 2). This involves tasks like modulation, coding, channel access, and error correction.

Resource Allocation

The eNB allocates radio resources, such as frequency bands, time slots, and physical channels, to UEs based on their communication needs and Quality of Service (QoS) requirements. It manages resource scheduling to optimize network efficiency.

QoS Control

The eNB enforces QoS policies, ensuring that diverse types of traffic, such as voice, video, and data, receive the appropriate level of service. It prioritizes traffic and allocates resources accordingly to meet QoS targets.

Mobility Management

The eNB manages the mobility of UEs within its coverage area. It facilitates handovers between cells or eNBs to ensure that UEs experience seamless and uninterrupted connectivity as they move from one place to other within the coverage areas.

Cell Selection and Reselection

UEs periodically measure the signal strength of neighboring cells to connect the desirable eNB having the better RF signal. The eNB provides information to UEs to assist in cell selection and reselection decisions, helping UEs choose the best cell for optimal service.

Bearer Management

The eNB establishes and manages bearers, which are logical communication channels used to transmit user data between the UE and the core network. It can set up, modify, or release bearers based on user demand and network conditions.

CP Functions

The eNB manages control plane signaling, including connection establishment, release, and maintenance. It communicates with Mobility Management Entity (MME) and Serving Gateway (SGW) in the EPC to establish and manage UE connections.

UP Functions

In the user plane, the eNB forwards user data packets between the UE and the core network. It encapsulates, segments, and routes data packets as needed to ensure that user traffic reaches its destination.

Security

The eNB plays a role in LTE security by implementing security mechanisms such as encryption and authentication for user data and control signaling. It helps protect against eavesdropping and unauthorized access.

Network Management and Monitoring

The eNB provides real-time monitoring and management capabilities, allowing NPSBN operation team to monitor the performance of the cell, diagnose issues, and optimize network parameters for improved efficiency.

Interoperability

LTE networks often need to interoperate with other cellular technologies and radio systems (e.g., non-LTE networks – Land Mobile Radio). The eNB may support interworking and handovers to ensure seamless communication across different networks.

6.1.1 Enhanced Cell Global Identity and Cell Configuration

An Enhanced Cell Global Identity (eCGI) is the 3GPP specifications (7) used globally to identify a specific cell tower. An eCGI consists of:

1. Mobile Carrier Code (MCC): An MCC is a three-digit number (XXX) uniquely assigned to each mobile carrier globally (8) except for the following countries which include a range of MC.
   - Values 310 through 316 (United States)
   - Values 404 through 406 (India)
   - Values 440 through 441 (Japan)
   - Values 460 through 461 (China)
   - Values 234 through 235 (United Kingdom)

   The MCC is used within the International Mobile Subscriber Identity (IMSI) to uniquely identify mobile subscribers and within a Location Area Identification (LAI) to uniquely identify a location area within a mobile network.
2. Mobile Number Code (MNC): An MNC is a two or three-digit number uniquely assigned within a country to a mobile carrier.
3. Cell identity (CellID): A 28-bit number uniquely assigned to a cell tower (9). A cell Id consists of eNBID which contains specific provider market and state number and cell number.

It should be noted that "s1b1" which stands for service 1, Broadcast 1, is used for emergency alert broadcast. All NPSBN towers have the capability to broadcast emergency alerts.

Table 6.1 provides the composition of an eCGI which include MCC and MNC values for FirstNet.

TABLE 6.1

Composition of an ECGI and PLMN Assignment

MCC	MNC	CellID
313	100	eNBID+Cell Number

TABLE 6.2

NPSBN eCGI Composition and Format

20 bits Cell Identity (CellID Range 0–268435455)	
eNB ID (eNB ID range 0–1048575)	Cell number (Cell no. range 0–55)
20 bits (3GPP 36-401)	8 bits

The combination of MCC and MNC is called Public Land Mobile Network (PLMN) which uniquely identifies a specific mobile operator in a country. The PLMN for NPSBN is 310-100 where MCC is 310 and MNC is 100.

Table 6.2 provides an eCGI composition format.

It contains the two components of an LTE CellID which are eNB-ID and cell number. eNB-ID is 20 bits long and has a value between 0 to 1048575 which also includes a two-digit market/state number which is a mapping of market/state code (e.g., 12 for VA and 42 for WA). An eNB-ID, technically called PLMN, must be unique in the NPSBN. The cell number is 8 bits long and has a value between 0 to 255.The combination of eNB-ID and cell number is called cell identity which is globally unique.

PLMN (MCC+MNC) is a significant feature to provide access to a specific LTE network provider's mobile network and to establish communication over a cell tower.

6.1.2 NPSBN and Cell Tower Installation and Sharing

In the early days of mobile network deployment, wireless network providers generally started with installing their own dedicated cell towers for their use. However, installing a RAN tower is extremely time consuming, expensive, and involved legal framework for land acquisition. Nowadays there are many private companies who have erected cell towers and continuously adding new sites and lease the space to various service providers resulting in a significant cost reduction for installation and operations for the shared towers.

6.1.3 Cell Tower Triangulation

One of the noteworthy features of the LTE eNBs is the ability to find the approximate location of the UE while it is communicating with a cell tower by measuring the signal time delay. However, the UE location accuracy increases when the radio signals from two or more towers are included for location computation using triangulation techniques. Generally, using cell tower triangulation having three towers signals computation, and the UE location can be found within an area of "about" 1/2 square mile.

The GPS provides better location accuracy compared to triangulation approximately within 5–10 feet. GPS does not require cell triangulation, and the UE is tracked using satellite signals. The GPS feature is now an integral part of the newer cell phones. It is also worth mentioning here that in the United States, e9-1-1 emergency systems can locate a UE interacting with a 9-1-1 emergency dispatcher accurately using the Location-Based Service (LBS).

6.1.4 Multi-Operator Core Network

The 3GPP Multi-Operator Core Network (MOCN) architecture standards (10) enable the operator to share the same RAN among more than one PLMN and multiple EPCs. No information concerning the configuration of a shared network is indicated to the UE.

MOCN-enabled eNB is shared by multiple operators; the system information broadcasted in each shared cell contains the PLMN-ID of each operator (up to 6 operators) and a tracking area code (TAC) valid within all the PLMNs sharing the radio access network resources. On the basis of system broadcast, UE selects desired PLMN and sends it to eNB during RRC connection setup. The eNB includes the selected PLMN in the "INITIAL UE MESSAGE" and forwards it to MME. The MME verifies and indicates the selected core network operator PLMN-ID to the UE in the Globally Unique Temporary Identifier (GUTI) during initial attach procedure.

Figures 6.3 and 6.4 highlight NPSBN MOCN feature enabling sharing the RAN between FN and AT&T subscribers. FN subscribers access the FN core whereas AT&T subscribers attach to AT&T core, respectively.

The key features of the MOCN are as follows:

- MOCN allows up to six separate core networks (or six separate operators) to be connected to the same RAN, for example, RAN will be connected to both FN core and AT&T core.

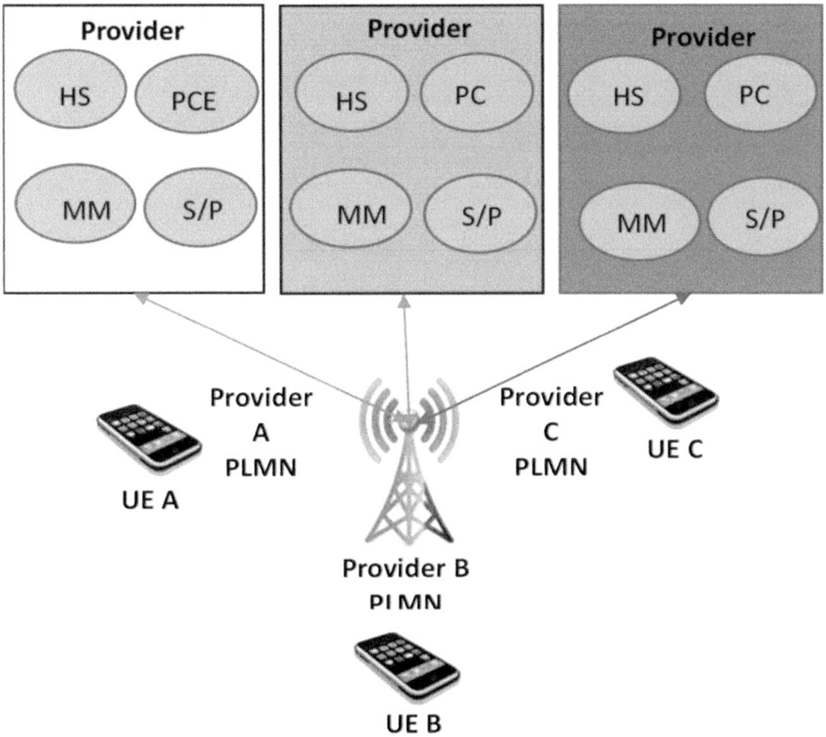

FIGURE 6.3
3GPP MOCN Architecture.

- The shared RAN will broadcast multiple Mobile Country Code (MCC) and Mobile Network Code (MNC) in the Radio Channel (SIB1), i.e., FirstNet PLMN 313-100 and AT&T PLMN 310-410.
- MOCN is an important feature when the network moves closer towards software-defined network and edge-computing becomes the norm for 5G.
- MOCN is used for clear traffic separation and introduction of specific features for a particular user group (such as FN) from EPC perspective without the need for dedicated RAN deployment.

The PLMN is categorized in the following ways:

- Primary PLMN is the main PLMN configured on eNB level, which is the same for all the cells in the eNB.

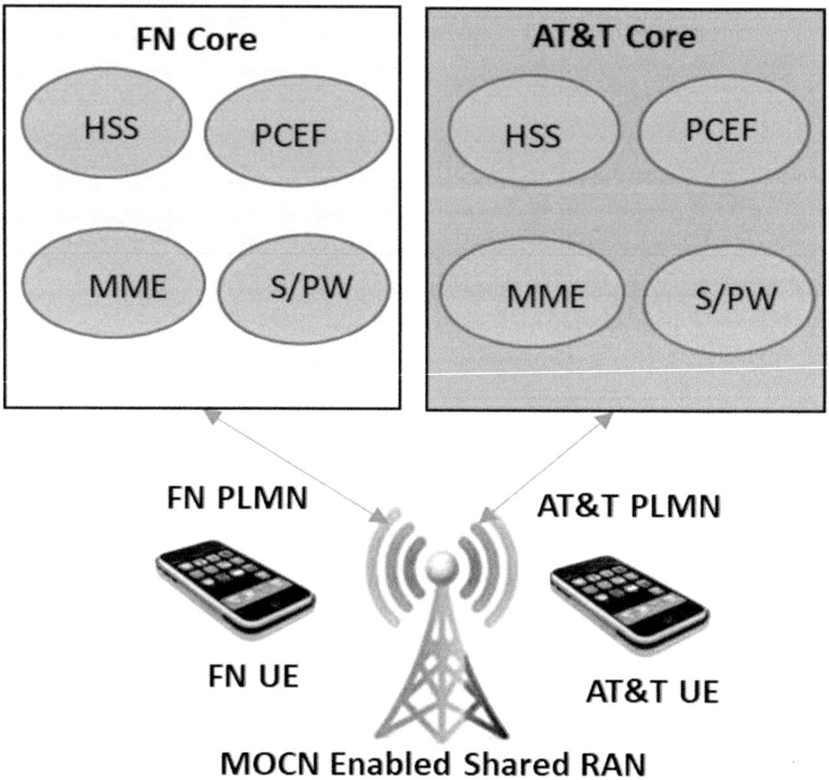

FIGURE 6.4
NPSBN MOCN PLMN Assignment.

- Serving PLMN: Serving PLMN is the PLMN that the MME has selected to serve the UE. Normally, it is the same PLMN that the UE selects when it connects to the network.
- Active PLMNs: Active PLMNs in a cell are all configured PLMNs in the cell, which are served by at least one MME.
- Equivalent PLMNs: Equivalent PLMNs are a list of PLMNs stored in the UE that are considered as equivalent as the serving PLMN for PLMN selection and cell selection/reselection of the UE.
- Configured PLMNs: Configured PLMNs are all PLMNs that an eNB can use and they are configured by operator. Configured PLMNs consist of a primary PLMN and a list of additional PLMNs.
- Additional PLMNs are a list of PLMNs that the operators can configure for each cell.
- Registered PLMN: Registered PLMN is the PLMN on which the UE has successfully completed location registration. In a shared network, the

TABLE 6.3

PLMN Assignment in the U.S. Mainland and Territories

Service Location	Wireless Service Provider	PLMN Assignments	Subscribers
U.S. States	AT&T	310-410 Primary (AT&T PLMN	AT&T retail and business subscribers
		313-100 Other PLMN (FN Primary PLMN)	First responders
Puerto Rico/U.S. Virgin Islands	LLA	313-790 Primary PLMN (LLA PLMN)	LLA retail and business subscribers
		313-100 Other PLMN (FN Primary PLMN)	First responders
Guam and CNMI	DOCOMO	310-470 as the Equivalent PLMN for 313-110	DOCOMO retail and business subscribers
		313-110 (FN Primary PLMN)	First responders
American Samoa	ASTCA	311-780	ASTCA retail and business subscribers
		313-100 (FN Primary PLMN)	First responders

registered PLMN is the PLMN defined by the PLMN identity of the core network operator that has accepted the location registration.

6.1.5 PLMNs and Primary PLMN Assignments

NPSBN has been assigned PLMN 313-100 to provide broadband network to first responders in all 50 states, Puerto Rico, and the U.S. Virgin Islands. AT&T has its own PLMN 310-410 to serve its retail and commercial subscribers. Using the MOCN features, AT&T successfully shares all bands including B14 among first responders and its own subscribers having PLMN 313-100 and PLMN 310-410, respectively.

Table 6.3 provides the MCC-MNC and the primary PLMN allocated for FirstNet users to AT&T and service providers in the U.S. territories.

6.2 FIFTH GENERATION (5G) NETWORK LOGICAL ARCHITECTURE

The Third Generation Partnership Project (3GPP) introduced LTE, a global standard for 4G wireless data transmission, in 2008 to meet the demand of

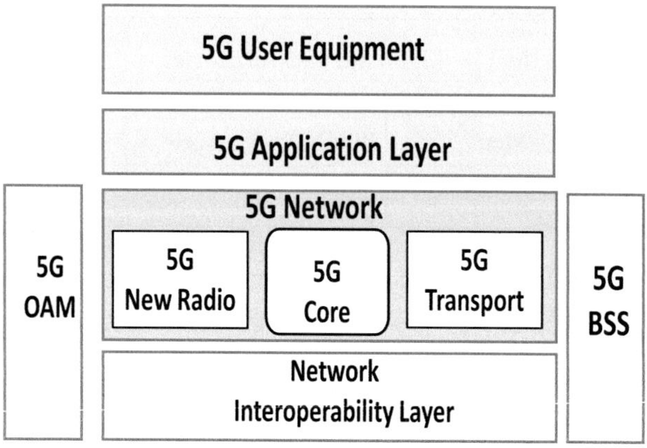

FIGURE 6.5
5G Network High-Level Logical Architecture.

huge amount of data at a faster data transmission rate compared to 3G. Since the adaption of 4G globally, development of standardization of intensive applications in all business domains have exploded, e.g., emergency services, Internet of Things (IOT), and interactive gaming, creating immense needs for faster data transmission, higher bandwidth, data security, and QoS.

Advent of 5G network initiated by 3GPP officially in 2015 focused on the requirements and architecture for new generation of mobile networks which includes 5G New Radio (5G NR), 5G Core Network (5G CN), 5G-capable devices, and various associated subinfrastructure. Figure 6.5 provides a high level and simplified logical architecture of 5G mobile network.

6.2.1 5G RAN/gNB

The introduction of 5G is to address many of the issues currently faced in 3G and 4G networks and incorporating new features, e.g., Multi-Input-Multi-Output (MIMO), beamforming, network slicing, virtualization, and cloud RAN.

MIMO utilizes many antennas at the 5G RAN base unit to improve capacity and coverage through spatial multiplexing. The beamforming feature directs radio signal in a specific direction to improve signal quality and reduce interference. Network slicing feature enables creating multiple virtual networks on a single physical infrastructure. Each slice can be customized for specific uses.

5G RAN is generally referred to as NR or gNB as well (see Figure 6.6).

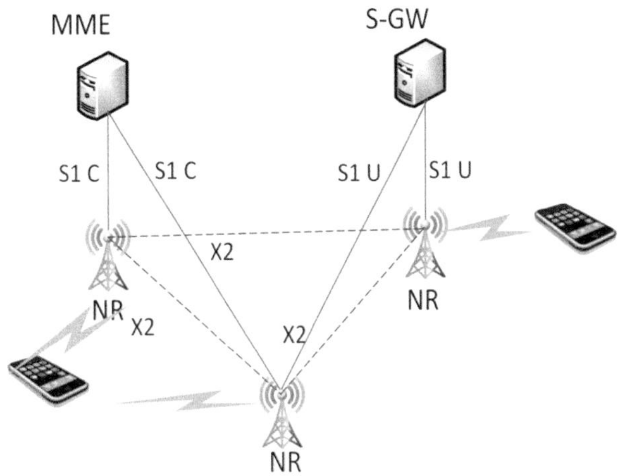

FIGURE 6.6
5G Network – NR Connectivity.

6.2.1.1 5G Spectrum

5G spectrum frequencies are generally classified into low band, mid band and high band, each providing different coverage areas, transmission speeds, transmission latencies, and specific situations. Table 6.4 provides various characteristics of the 5G spectrum bands.

In addition, millimeter Wave (mmWave) refers to higher frequency radio bands ranging from 24 GHz to 40 GHz. Similarly, sub-6 GHz refers to mid and low frequency bands under 6 GHz. The mmWave 5G offers greater bandwidth, relieving network congestion. In crowded areas, LTE speeds can be slow because of the number of devices connecting, while mmWave feature is able to manage a greater number of connections without any significant speed drops. Ideally, mmWave is set up in crowded areas where network congestion is a problem, such as at sporting events, airports, concerts, and other locations where many people gather, as well as in urban areas. It is apparent that full-scale deployment of 5G globally will support varieties of situations to meet specific requirements.

6.2.1.2 5G RAN Architecture – Control Plane and User Plane Separation (CUPS)

One of the fundamental enhancements in 5G RAN is the separation of the control and user plane signals and creating two new interfaces to connect

TABLE 6.4

5G Radio Spectrum

Bands	Radio Frequencies Range	Common Bands and Coverage	Spectrum Characteristics
Low band	Less than 1 GHz	600 MHz (n71) 700 MHz (n28)	Low bands provide broad areas of coverage and good penetration through buildings and obstacles but lower data speed. This band provides an ideal situation for rural and suburban areas requiring wide coverage. Initially, low bands were used in 3G which many U.S. operators are turning off and reframing to use for 5G.
Mid band (also called sub-6)	Range: 1GHz–6 GHz	2.5 GHZ (n41), 3.5 GH (n78, n77), 4.5GHz (n79)	Mid bands provide a spectrum where both coverage and speed are balanced to meet the transmission requirements. Suitable for suburban and urban areas requiring both coverage and speed. Many countries have already designated 2.5 GHz–4.5 GHz a 5G spectrum.
High band (also called millimeter Wave or mmWave)	More than 6 GHz mmWave – 24 GHz–40 GHz	24 GHz and 100 GHz) 26 GHz (n258), 28 GHz (n257), 39 GHz (n260)	High bands transmit very high data speed but relatively very low area coverage and poor penetration through building, folios, etc. Most ideal for densely populated areas including airports, stadiums, etc. Signal travels less than a mile's distance.

MME and SGW. The architecture and functionality for control and user plane separation of SGW, PGW, and TDF is based on the following concepts:

- Interworking with networks not applying control plane (CP) and user plane (UP) separation is possible (e.g., in case of roaming scenarios)
- 5G split network entities can interwork with network entities that are not split within the same network
- 5G split network entities have no requirement to update UE and RAN
- The SGW/PGW selection function of the MME, ePDG, and Trusted Wireless Access Network (TWAN) described in TS 23.401 (11) and TS 23.402 (12) is used for the selection of the respective CP function

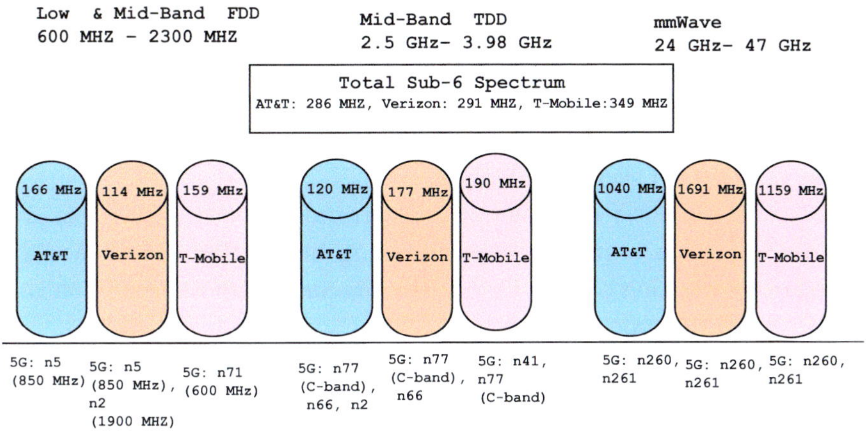

FIGURE 6.7
FCC Spectrum License to Major Service Providers (as of 2022).

- The configuration-based mechanism (in PGW or PCRF) is used for the selection of the CP function
- A CP function can interface with one or more UP functions to enable independent scalability of CP functions and UP functions

The components of the 5G core architecture include Mobility Management Function (AMF), UP Function, and Data Network (DN) Core Access. DN includes operator services, internet access, and third party services.

The gNB nodes support NR devices via the NR CP and UP protocols. A gNB is a logical node rather than a purely physical one and is commonly deployed as a three-sector site.

The Traffic Detection Function (TDF) is a functional entity that performs application detection and reporting of detected applications and its service data flow description to the PCRF. The PCRF encompasses service data flow detection, policy enforcement, and flow-based charging functionalities. (13)

Layer 3 in networking refers to the network layer in the Open Systems Interconnection (OSI) model. This layer is responsible for routing data packets between devices across different networks. It uses logical addressing (like IP addresses) to identify devices and ensure that packets are sent to their correct destinations. Common protocols that operate at Layer 3 include Internet Protocol (IP), Internet Control Message Protocol (ICMP), and Internet Group Management Protocol (IGMP).

6.2.2 5G NR Layers

5G NR consists of several layers that facilitate communication between the UE and the network. These layers are primarily divided into:

Layer 1: Physical Layer (PHY)
This layer is responsible for the transmission and reception of raw data over the air and other critical features including modulation and demodulation, error correction coding, Orthogonal Frequency-Division Multiplexing (OFDM), MIMO processing, and beamforming.

This channel is the pathway for data as it moves across the physical layer. It defines how data is processed and delivered to the higher layers and include channels such as the Broadcast Channel (BCH), Downlink Shared Channel (DL-SCH), and Uplink Shared Channel (UL-SCH).

These layers work together to optimize data transmission, improve efficiency, and enhance user experience in mobile communication.

Physical Layer Procedures: Handles various procedures like cell search, power control, synchronization, random access, HARQ (error correction), beam management, side link communication, and channel access. Implicitly supports interference coordination through resource control.

Physical Layer Measurements: UE and network measure radio characteristics (e.g., signal strength, timing) and report them to higher layers for tasks like handover and resource management.

Layer 2: Data Layer
This layer can be subdivided into three sublayers:

- MAC: Manages protocol data units (PDUs) and handles multiplexing onto the physical channels, scheduling, error correction through Hybrid Automatic Repeat Request (HARQ), and dynamic allocation of radio resources.
- Radio Link Control (RLC): Ensures the reliable delivery of data. It is responsible for segmenting and reassembly of long Service Data Units (SDUs) and providing error correction through Automatic Repeat Request (ARQ).

- Packet Data Convergence Protocol (PDCP): Compresses and decompresses the headers of IP packets and performs encryption and data integrity protection.

Layer 3: Network Layer

This layer can further be subdivided as follows:

- Radio Resource Control (RRC): This sublayer manages the control plane signaling between the UE and the network, including the configuration of the lower layers, system information broadcasting, session management, and mobility management.
- Multiple Access: Uses OFDM with a cyclic prefix (CP) for efficient data transmission. Supports Discrete Fourier Transform-spread OFDM (DFT-s-OFDM) for uplink. Employs both Frequency Division Duplex (FDD) and Time Division Duplex (TDD) for paired and unpaired spectrum access.
- Resource Blocks: Uses resource blocks, each spanning 12 subcarriers, as the basic unit for resource allocation. This agnostic approach adapts to various spectrum allocations.
- Frame Structure: Radio frame has a duration of 10 milliseconds (ms) and consists of 10 subframes (each 1ms). Subframes are further divided into slots (each with 14 symbols).
- Physical Channels: Downlink channels: PDSCH (shared), PDCCH (control), PBCH (broadcast). Uplink channels: PRACH (random access), PUSCH (shared), PUCCH (control). Sidelink channels: PSBCH (broadcast), PSCCH (control), PSFCH (feedback), PSSCH (shared).
- Modulation: Downlink: QPSK, 16QAM, 64QAM, 256QAM, 1024QAM; Uplink: OFDM with CP: QPSK, 16QAM, 64QAM, 256QAM. DFT-s-OFDM with CP: $\pi/2$-BPSK, QPSK, 16QAM, 64QAM, 256QAM.
- Channel Coding: Uses quasi-cyclic LDPC codes for data channels (PDSCH, PUSCH). Employs polar coding for control channels (PDCCH, PUCCH, PBCH).

6.3 SMALL CELL AND DISTRIBUTED ANTENNA SYSTEM

Like LTE macro cells (eNBs), small cells are a type of cellular-based station in small sizes and provide small coverage compared to regular eNBs. These are low-powered radio access nodes and support coverage, capacity, and connectivity where macro cells are not reachable or radio signals are fairly poor, where installing macro cells would not be economically nonsustainable, and in indoor and highly congested areas.

Small cells are classified into microcells, picocells, femtocells, and metro cells. Table 6.5 provides the basic features.

Typically, the small cells vendors (e.g., Ericsson, Nokia, Corning/Spider Cloud, Commscope) currently do not support FirstNet B14. However, AT&T, contractually responsible for NPSBN, is closely collaborating with vendors to support MOCN features enabling first responders to access their devices and QPP support over small cells.

Recently, Corning/Spider has announced the 5G-enabled small cells ready for deployment. Corning stated that 5G cellular services are easily layered onto the LTE E-RAN with a simple software update, and installation of 5G E-RAN system in the building is feasible. It is realistic to state that most of the vendors will start supporting 5G small cells in the near future.

TABLE 6.5

Types of Small Cells and Their Characteristics

Type of Small Cell	Coverage Range	Power Output	Uses
Microcells	Hundreds of meters to a few kilometers	1–10 watts	To enhance coverage and capacity in urban areas, supplement coverage gaps, and handle high-volume traffic.
Picocells	Tens to hundreds of meters	Less than 1 watt	Provide coverage in indoor areas, e.g., office buildings, airports, shopping malls.
Femtocells	Tens of meters	Less than 100 milliwatt	Enhance coverage in residential homes and small offices, and are often connected via users' broadband connections.
Metro Cells	Typically, about 100–200 meters	1–5 watts	Enhance coverage and capacity and most suited for high-traffic areas, city centers, and are often installed in lampposts, buildings, etc.

Cellular Distributed Antenna System (DAS) is a network of spatially separated antenna nodes that provide wireless service within a building or facility. Cellular DAS can be configured for major cellular carrier frequency bands and those used in private cellular networking (CBRS). In general, DAS is suited for in-building facilities having very poor coverage or no coverage at all from the macro cells and requires connectivity to significantly large number of users such as

- Stadiums and arenas
- High-rise commercial buildings
- Healthcare complexes
- Educational institutions
- Malls and retail centers
- Manufacturing facilities and warehouses
- Hotels and convention centers

6.4 NPSBN RAN – KEY POINTS

NPSBN uses RAN that connects individual FN devices to the FN core network enabling wireless communications. A RAN consists of radio antenna, RRH, and BBU and their interconnections, allowing mobile users to access FN services such as voice, data, and video transmission. The antenna transmits/receives RF signals over the air interfaces with mobile devices wirelessly. RRH converts the RF signal into data signal and vice versa. BBU connects the LTE EPC core and manages the whole base station including signal processing and operational maintenance.

4G and 5G RANs have different radio spectrums though both can be deployed on the same cell tower. Generally, 4G and 5G RAN are also called eNodeB (eNB) and gNodeB (gNB), respectively. Wireless devices (phones, laptops, smart watches, etc.) can access 4G or 5G RAN or both depending upon their configuration by the service provider. Each cell tower has a unique global identity, called eCGI which consists of a three-digit MCC, 3-digit MNC, and 8 bits cell Id. FirstNet MCC+MNC is 313+100 and is also called PLMN. PLMN is part of the IMSI which uniquely identifies mobile subscribers globally.

MOCN is an enhanced feature enabling multiple service providers to use the same RAN. The MOCN enabled FN subscribers to share the RAN and radio spectrum with AT&T subscribers still accessing the FN dedicated core network.

FN subscribers also have access to small cells which provide the communication ability inside buildings or closed areas where metro cells do not have coverage or have a weak signal.

7

NPSBN Core Network

National Public Safety Broadband Network (NPSBN) is developed based on architecture and framework developed by many standards bodies, consortiums, and private organizations including 3GPP, ESTI, ISO, ITU, and IEEE. Architecture refers to the high-level conceptual structure and design principles of a system, defining its components, their relationships, and how they interact. An architecture provides a blueprint for the system's overall organization and design. On the contrary, a framework is a specific set of tools, libraries, and conventions that provide a foundation for network and developing applications within that architectural context. Frameworks often facilitate rapid development by providing reusable components and predefined functionality.

NPSBN core is the brain of the NPSBN network performing critical functions which are generally classified into control plane and user plane functions. Control plane basically deals with signal processing, communications with user device over eNB, user authentications, attachment process, etc., whereas the user plane facilitates user data packet transmission between multiple user devices and the packet data network (PDN).

The sections that follow describe these functions in detail.

7.1 LTE CORE/EVOLVED PACKET CORE

The LTE core, also known as Evolved Packet Core (EPC), is a crucial subsystem in the LTE network. EPC serves as the backbone of the LTE network. It is responsible for many key functions that enable the delivery of

DOI: 10.1201/9781003557739-9

high-speed multimedia services to mobile users. The following are the significant functions of the EPC.

Authentication and Security: The EPC handles authentication and authorization of user devices. It verifies the identity of devices and grants them access to the network based on user credentials and security keys. It supports data security through encryption and integrity protection.

Mobility Management: The EPC manages the mobility of user devices as they move within the LTE network. It tracks the location of devices, facilitates handovers between eNBs, and ensures seamless connectivity during mobility events.

Network Gateway Functions: The EPC includes components like the Serving Gateway (SGW) and Packet Data Network Gateway (PGW). The SGW routes user data packets within the LTE network, while the PGW acts as the interface between the LTE network and PDNs.

Policy Control and Enforcement: The EPC enforces policies defined by network operators to control data usage, manage network congestion, and optimize resource allocation.

Quality of Service (QoS) Management: One of the functions of the EPC is to enforce QoS policies and prioritize diverse types of traffic. This ensures that critical services like voice calls and real-time videos are given higher priority to maintain a consistent user experience.

Charging and Billing: The EPC keeps track of data usage for billing purposes. It monitors the volume of data transmitted by each user device and generates records for billing systems.

Packet Routing and Forwarding: The EPC routes and forwards IP packets between user devices and external networks. It determines the most efficient path for data transmission and ensures that data packets reach their intended destinations.

7.1.1 FN NPSBN 4G Dedicated Core

As stated in Section 2.1. Middle Class Tax Relief Act of 2012 authorized FirstNet Authority (FNA) to establish an NPSBN based on a single, national network architecture to build a dedicated, reliable wireless network for first responders in all U.S. states and its territories, including rural communities and tribal nations. On the basis of the NPSBN requirements, a dedicated FN LTE core was built to serve first responders.

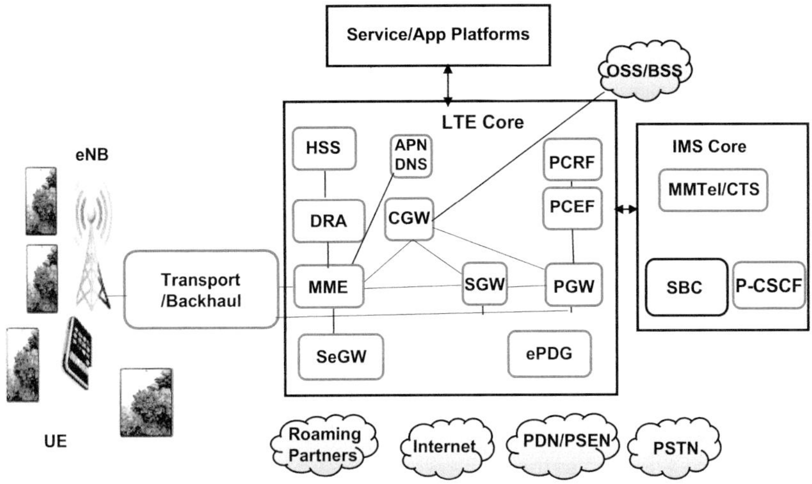

FIGURE 7.1
NPBN-Dedicated LTE Core Architecture.

A dedicated FN LTE core is accessible only to FN-enabled devices and cannot be accessed or attached by any other, including American Telephone and Telegraph Company (AT&T) customers with their own non-FN devices.

Figure 7.1 illustrates a high-level logical architecture of NPSBN EPC. Here is the brief description of the basic components of the 4G core.

7.1.1.1 Mobility Management Entity

The Mobility Management Entity (MME) is a crucial component of the EPC network architecture. The MME acts as the primary control point for managing the mobility of subscribers within the network and serves as a control point for authentication of mobile subscribers within the network. The MME manages the control plane signaling between the user equipment (UE) and the core network, and coordinates user plane signaling with the SGW based on the subscriber's location. When a mobile device attaches to the network, the MME authenticates the subscriber, authorizes access to network resources, completes the attachment process, facilitates packet transfer, QoS, subscriber's mobility, roaming, etc.

The MME communicates with other network elements using various protocols, including the diameter protocol for authentication and authorization, the S1-MME protocol for control plane signaling between the UE and the MME, and the S11 protocol for communication between the MME and the SGW/PGW.

The MME performs several key functions within the EPC network, which include:

Authentication: When a subscriber attaches to the network, the MME is responsible for authenticating the subscriber's identity. The MME verifies the subscriber's credentials and authorizes access to the network resources. If the subscriber's credentials are invalid or unauthorized, the MME will reject the attached request.

Mobility Management: The MME is responsible for managing the mobility of subscribers within the network. It manages the tracking area updates, which enable the network to know the location of the UE. The MME is also responsible for controlling the handover of subscribers from one eNB to another using X interface, ensuring that the subscriber's connectivity is maintained during the transition.

Security: The MME is responsible for ensuring the security of the network and the subscriber's data. It provides secure communication between the UE and the network by setting up security associations and encryption keys. The MME also manages the security context between the UE and the SGW/PGW, ensuring that the data exchanged between them is secure.

QoS Management: The MME is responsible for managing the QoS for the subscribers within the network. It assigns QoS parameters to data sessions based on the subscriber's subscription and network conditions. The MME also manages the handover of subscribers between different Radio Access Technologies (RATs), ensuring that the subscriber's QoS requirements are maintained during the transition.

Paging: The MME is responsible for paging subscribers within the network. When a data packet arrives at the SGW/PGW, the MME determines the UE's location and initiates a paging request to the appropriate base station to locate the UE.

Roaming: The MME is responsible for managing the mobility of the subscribers when they roam outside their home network. It communicates with the Home Subscriber Server (HSS) to retrieve subscriber information and authentication credentials. The MME also communicates with the visited network's MME to establish connectivity and ensure that the subscriber's QoS requirements are maintained.

The architecture of the MME is designed to be highly available and scalable, ensuring that it can handle many subscribers and network traffic. The MME is typically deployed in a redundant configuration to ensure high availability and fault tolerance. The MME can also be virtualized to reduce the hardware footprint and enable flexible deployment options.

7.1.1.2 LTE EPC Functional Design Interfaces

Figure 7.2 presents the main interfaces among various EPC nodes. These interfaces are specified by 3GPP LTE core specifications.

7.1.1.3 System Architecture Evolution Gateway

System Architecture Evolution Gateway (SAE-GW) elements provide SGW and PDN Gateway (PGW) functionality within the NPSBN EPC.

SGW: The SGW routes and forwards user data packets, while also acting as the mobility anchor for the user plane during inter-eNodeB handovers and roaming. For idle state of UEs, the SGW terminates the downlink data path and triggers paging when downlink data arrives for the UE.

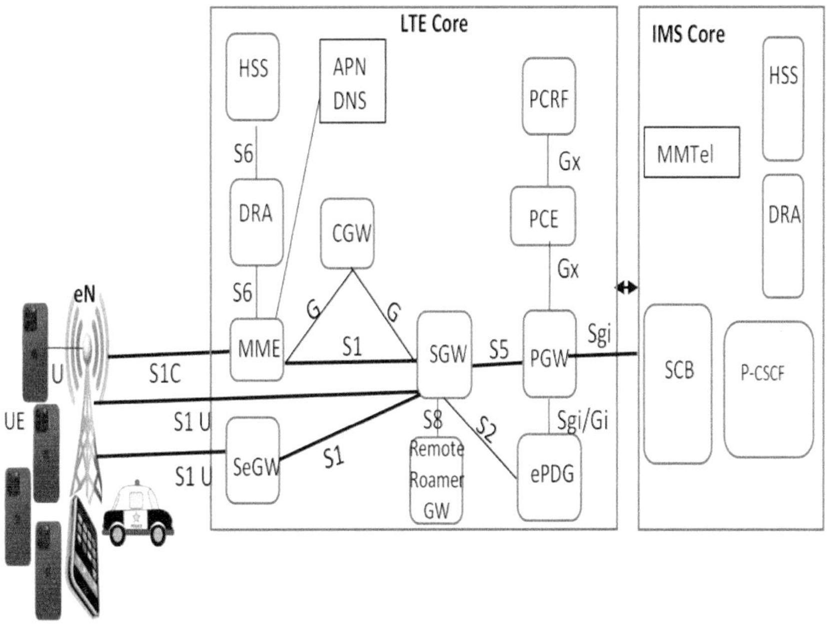

FIGURE 7.2
NPSBN EPC Interfaces.

It manages and stores UE contexts, e.g., parameters of the IP bearer service and network internal routing information. It also performs replication of the user traffic in lawful interception.

PGW: PDN-GW plays a crucial role in managing and routing data traffic between the LTE network and external PDNs, such as the internet and private corporate networks. Here are its main functions:

The PDN-GW provides connectivity from the UE to external PDNs by being the point of exit and entry of traffic for the UE. A UE may have simultaneous connectivity with more than one PGW for accessing multiple PDNs. The PGW performs policy enforcement, packet filtering for each user, charging support, lawful interception, and packet screening. Another key role of the PGW is to function as the anchor for mobility between 3GPP and non-3GPP technologies such as WiMAX and 3GPP2 (CDMA 1X and EvDO).

Following are the critical functionalities of an NPSBN PGW:

Traffic Routing and Forwarding: The PDN-GW routes and forwards data packets between the LTE network and external networks, ensuring that user data is transmitted correctly.

QoS Management: The PDN-GW classifies and prioritizes data traffic based on QoS parameters to ensure diverse types of traffic receive the appropriate level of service.

Policy Enforcement: It enforces network policies, including traffic shaping, bandwidth allocation, and security policies.

Network Address Translation (NAT): The PDN-GW performs NAT to allow multiple devices within a private LTE network to share a single public IP address.

Packet Filtering and Security: It performs packet filtering and firewall functions to protect the LTE network from malicious traffic.

Charging and Billing: The PDN-GW keeps track of data usage by individual UEs for accurate billing.

Mobility Management: It ensures that data sessions can be maintained as UE moves within the LTE network.

Gateway to Different Networks: The PDN-GW connects to various types of external networks, including the public internet and private corporate networks.

NPSBN SAE-GW Selection Process for FirstNet Users: In NPSBN LTE network, the Domain Name System (DNS) is used to manage the selection of the PDN-GW, SGW, MME, and HSS within the NPSBN. During the UE attachment process, the eNB assigns MME based on configured MME pool. The MME queries the Access Point Name (APN) DNS server to select the SGW and appropriate PDN-GW where a requested (subscribed) PDN connectivity (APN) is located. Selection can be based on the information provided to the MME, when the UE attaches to the network.

NPSBN core physical deployment locations are crucial to meet the availability, accessibility, performance, and latency expectations and have both logical and physical redundancies to meet any disaster situation. To meet these critical requirements, the United States is categorized into three zones and two NPSBN core centers are deployed in each zone.

East Zone:Northeast – New York (Broadway), and Southeast: Alpharetta, GA
Mid-West Zone:North Central – Akron, OH, and South Central: Allen, TX
West Zone:Northwest – Bothel, WA, and Southwest – Concord, CA

Each zone serves as a primary core for the first responders located in the zone. The other two zones would serve as secondary and tertiary and provide access in case both centers in primary zones are down.

The FirstNet PGW will be enhanced to support Additional Protocol Configuration Options (APCO)-based discovery of the primary and secondary P-CSCF IPv6 addresses on initial attach on Wi-Fi. This is equivalent to the PCO-based P-CSCF discovery mechanism defined for VoLTE. APCO within Create Session Request from the ePDG contains a request for P-CSCF IPv6 address. The PGW selects the next available primary and secondary P-CSCF IPv6 addresses from the existing APN-specific P-CSCF selection table currently used for IMS services in Radio Access Network (RAN). These selected addresses are returned to the ePDG in APCO in Create Session Response.

Table 7.1 illustrates the APN and Tracking Area ID (TAI) queries based on various mobility scenarios that a FirstNet subscriber can encounter in NPSBN.

TABLE 7.1

NPSBN APN and TAI

Scenarios	Serving-PLMN	MME	APN DNS	APN (PGW) Query	TAI (SWG) Query
LTE attach: 313-100 PLMN available	313-100	FN MME	FN APN DNS	mnc100.mcc313	mnc100.mcc313

TABLE 7.2

FirstNet APN Used in NPSBN EPC

APN	Description	APN Code
Custom APN	An APN name unique to the specific FirstNet customer, used for data services. If the FirstNet customer requires VPN connectivity to reach their enterprise network, a customer APN must be used. However, a customer may also use a custom APN to reach AT&T services or the public internet.	<custom apn>
FirstNet Phone	A general-purpose APN for FirstNet used for smartphones. This APN is used to reach internal AT&T services (e.g., UT flows supporting VoLTE) and the public internet.	firstnet_phone
FirstNet Broadband	A general-purpose APN for FirstNet used for data-only devices (e.g., tablets, laptops). This APN is used to reach internal AT&T services and the public internet.	firstnet_broadband
Nxtgenphone	General data for legacy devices	
IMS	APN for IMS services, such as VoLTE, IR, IPME.	ims
SOS	APN for E9-1-1 emergency service	sos
Hotspot	Tethering for legacy devices	Hotspot
MC PTT	Mission Critical Push To Talk	mcptt apn

7.1.1.4 APN DNS

APN is the code that facilitates the FirstNet UE access to desired data services (PDN) and internet while managing user authentication and authorization. FirstNet has specified a set of six APNs for the use of first responder accessing NPSBN.

NPSBN will have its own dedicated APN DNS services. APN DNS will provide name resolution including FirstNet enterprise APN and custom APNs as Public Safety Entities have their unique APNs. The AT&T commercial DNS will be populated with FirstNet records to ensure that the FirstNet subscriber sessions are directed to gateways in the NPSBN Core.

The current APN used for FirstNet users are listed in Table 7.2. It should be noted that additional APN will be added in the future as the need for new APN arises.

7.1.1.5 Home Subscriber Server (HSS) and Master Integrated Network Directory (MIND)

The NPSBN HSS is a central database that contains user-related and subscription-related information. The HSS database includes functionalities such as mobility management, call and session establishment support, user authentication, and access authorization.

HSS is also the master profile database of NPSBN subscribers managed by FN/AT&T. Besides managing a master list of features and services that are associated with each subscriber, and supporting access authentication and authorization of the user, the HSS provides subscriber profile information to EPC network components, such as the MME. The HSS provides the following:

- Subscriber profile
- Access authentication
- Features and services
- Mobility management

7.1.2 Master Integrated Network Directory (MIND)/ and Subscriber Profile Repository (SPR)

Master Integrated Network Directory (MIND) is an open-wave Lightweight Directory Access Protocol (LDAP) directory. MIND provides Subscriber Profile Repository (SPR) function to subscriber profiles for service validations and QoS policies, etc.

7.1.2.1 Diameter Routing Agent

A Diameter Routing Agent (DRA) is a functional element in a 4G network that provides real-time routing capabilities to ensure the messages are routed among the correct nodes in NPSBN. AT&T shares the EPC DRA infrastructure for NPSBN. The shared DRA forwards the NPSBN specific transactions to the appropriate NPSBN nodes. The DRA routes Public Service Entity (PSE) requests to the NPSBN HSS.

MMEs across all locations are regionally mapped with the closest DRA location. This design approach ensures to select an option that mitigates risk of traffic overload, allows efficient routing, and minimizes latency/round trip delay.

7.1.2.2 Policy and Charging Rules Function

Policy and Charging Rules Function (PCRF) is an EPC node responsible for determining policy rules in real time for each subscriber. It establishes the QoS and charging requirements for each requested service or bearer based on rules preconfigured with the subscriber's service profiles. These requirements are provided to the PGW which in turn ensures the rules are met. The following are some salient PCRF features for FirstNet.

It supports provisioned QoS settings defined by business rules for FirstNet subscribers in normal subscriber network attach and service requests. FirstNet subscribers normally have priority levels including Incident Uplift, Platinum, Gold, Silver, and Best Effort.

7.1.2.3 Support Up to Five Levels of Service, Priority, and Preemption Packages for FirstNet Subscribers

To support near-real-time subscriber QoS uplifting from Incident Management Portal and modify the QoS for existing active sessions for up to one thousand subscribers. The QoS uplifting applies to all users' active sessions and new sessions during the timeframe.

At the end of the uplifting period, PCRF must revert the QCI/ARP on impacted subscriber sessions to pre-incident levels upon receiving Rx notification when the next PCRF refresh timer expires; for MIND subscriber profile.

Policy and Charging Enforcement Function (PCEF).
The NPSBN PCEF element enforces QoS and other performance-related requirements specified in PCRF. PCEF interfaces with PGW and performs deep data packet inspection to ensure that each data packet is transmitted complying with the mandated PCRF requirements.

7.1.2.4 Evolved Packet Data Gateway

Evolved Packet Data Gateway (ePDG) provides interworking function between the NPSBN EPC and untrusted non-3GPP networks that require secure access such as a Voice over Wi-Fi (VoWiFi) and IR94 traffic offload (14).

In NPSBN, like many EPC elements, ePDG is cloud based and follows required redundancy design and multiple local deployments to support performance. EPDG instance selection is based on the proximity of the local

DNS versus the subscriber RAN location and about 95% of the time the nearest ePDG instances are selected.

To facilitate Wi-Fi calling, the following scenarios will be required:

- IMS-based services (IMS APN is "always on")
- VoIP calling in Wi-Fi (HD voice) with call continuity on handover between Wi-Fi and LTE
- E9-1-1 calling in Wi-Fi; E9-1-1 handover between Wi-Fi and LTE is out of scope
- Video calling in Wi-Fi with call continuity on handover between Wi-Fi and LTE
- SMS over IP
- IPME (chat, pager mode, large message mode) in Wi-Fi with session continuity on handover between Wi-Fi and LTE.

7.1.2.5 Charging Gateway

Charging Gate Way (CGW) interfaces with AT&T Business Support Systems (BSS) and provides user-specific data to be appropriately billed.

7.1.2.6 NPSBN IP Multimedia Subsystem

LTE IP Multimedia Subsystem (IMS) is a framework that allows multimedia services to be delivered over IP networks based on SIP/H.323/H.248 signaling developed by Internet Engineering Task Force (ITEF) (15) and International Telecommunication Union (ITU) (16). Session Broder Controller (SBC), Proxy Call Session Control Function (P-CSCF), and Corporate Telephone Number (CTN) nodes dedicated to FirstNet NPSBN and other IMS nodes are shared with AT&T subscribers. The NPSBN Partner, AT&T, has deployed Universal Service Platform (UPS) to provide IMS services.

Session Border Controller

A Session Border Controller (SBC) is an LTE network node that manages and protects IP communications, particularly in Voice over Internet Protocol (VoIP) networks. SBC key roles and functions include:

Security: SBCs protect the network from malicious attacks such as Denial of Service (DoS) attacks, toll fraud, and malformed packet protection. They also provide encryption for signaling and media.

Connectivity: SBCs enable various parts of the network to communicate by handling NAT traversal, SIP normalization, and protocol translations between different signaling protocols like SIP, H.323, and others.

QoS: SBCs manage the QoS policies of a network, including traffic policing, resource allocation, rate limiting, and call admission control.

Regulatory Compliance: SBCs support regulatory requirements such as emergency call prioritization and lawful interception.

Session Management: SBCs control the signaling and media traffic during the setup, conduct, and teardown of interactive media communications.

Proxy Call Session Control Function (P-CSCF)

A P-CSCF is an SIP proxy that is the first point of contact for UE in a mobile network. All SIP traffic to and from the user (17) handles signaling messages and acts as the ingress and egress point between the UE and the IMS domain, managing Session Initiation Protocol (SIP) traffic effectively.

Multimedia Telephony (MMTel)

Multimedia Telephony (MMTel) is a global standard developed by the Third-Generation Partnership Project (3GPP) and the European Telecommunications Standards Institute (ETSI) for providing multimedia communication services over IMS. Key features of MMTel are:

- *Converged Communication:* MMTel enables real-time multimedia communication, including voice, video, text, and file sharing, over both fixed and mobile networks.
- *Single Session Management:* It allows users to add or drop media during an ongoing session, such as starting with a chat and adding voice or video without ending the session.
- *Interoperability:* MMTel ensures interoperability between different operators, allowing users from different networks to communicate using the full set of media capabilities.
- *All-IP Network:* It uses IP transport over mobile access technologies like LTE and HSPA, making it future-proof and efficient.
- *Supplementary Services:* MMTel supports various supplementary services, such as conferencing, media sharing, and file transfer.

In NPSBN IMS subsystem, MMTel provides voice, video, and other telephony services over LTE network (VoLTE). MMTel uses IMS subsystem elements to deliver voice, video, and chat services to users. Along with that it specifies the way to share images, videos, and files in real time. CTS performs simultaneous ringing for incoming calls to the twinned devices.

7.1.3 FN 4G Shared Network Elements

The FNA has the tough task of identifying and selecting a private partner(s) capable of delivering NPSBN in stipulated timeframe and budget and has ability to monetize the network. Building the NPSBN from scratch was not feasible with the FNA budget allocated by the Congress. FNA considered for a partnership to build NPSBN in consultation with the state and federal governments (FEMA, Red Cross, etc.) and local public safety agencies in consistent with the vision of the Middle Class Tax Relief Act of 2012 to maintain long-term sustainability, innovation, and ability to serve public safety responders everywhere – urban, suburban, rural, tribal communities across 50 states and territories. FNA partnered with AT&T to develop and deploy NPSBN with the following salient points:

- The first responders have secure, uninterrupted, and priority network services to perform their day-to-day jobs for safeguarding lives and properties.
- NPSBN is monetized for its long-term sustainability.
- NPSBN is an innovative network and provides current and futuristic public safety related services.

In keeping with the above vision and constraints, NPSBN adapted the following design parameters:

- Shared radio access networks between FN and AT&T commercial users
- Shared transport network between FN and AT&T commercial users
- Dedicated 4G/5G core for FN users.
- Dedicated IMS subsystem for only a few elements; the rest of the elements are shared
- Shared Operation Support Systems (OSS) and BSS

This is an ideal design approach to meet the critical NPSBN requirements and budget constraints.

7.1.3.1 NPSBN LTE/4G Interfaces

3GPRS/LTE/4G defines several interfaces that connect different network components for efficient data flow. These interfaces are standardized by 3GPP to ensure interoperability and support seamless mobility and IP-based services.

For example, the **S1 interface** links the evolved NodeB (eNB) to the EPC (Evolved Packet Core), while the **X2 interface** connects eNBs for handovers and load balancing. Inside the EPC, interfaces like **S5/S8**, **S6a**, and **S11** coordinate signaling and bearer management across the MME, SGW, and PGW.

Table 7.3 provides a full description of the LTE interfaces.

7.1.4 LTE/4G Protocols, Interfaces, Controls, and Processes

NPSBN LTE/4G network comprises protocol stacks which specify rules for the interfaces and exchange of information between various mobility network

TABLE 7.3

LTE/4G EPC Interfaces

Reference Point	Interfaces	Description
UE-eNB	LTE-UU	Interface for the control and user planes between UE and eNB via air media
enB-eNB	X2	Interface for control and user planes between two neighboring eNBs. It is used during handover from one eNB to another eNB
eNB-MME	S1-MME	Interface for control plane between eNB and MME
eNB-SGW	S1-U	Interface for user plane between eNB and SGW for transmitting packet data
MME-SGW	S11	Interface between an MME and SGW; it provides a GTP tunnel per users
SGW-PGW	S5/S8	Interface for the control and user planes between SGW and PGW (PDN Gateway); it provides a GTP tunnel per bearer for the user plane and GTP tunnel management per user for control plane; for inter PLMN, S8 is used for roaming
MME-HSS	S6a	Interface between MME and HSS to exchange user subscription and authentication information
SGW-IMS	SGI	Interface between SGW and IMS core to exchange packet data based on PLMN
PCRF-MIND	Sp	Interface between PCRF and MIND to exchange subscriber profile
SGW- MSP, NAT, FW/PAT, FW Routing, DNS FW	SGI	Interface/reference point between SGW and different PDN to exchange packet data
SGW-OCG	Ga	Interface between SGW and OCG which serves the Call Detail Records (CDRs) which are written in S/PGW and sent to the OCG; this interface uses a GTP-based protocol, with modifications that support CDRs

layers and sublayers. The LTE network layer design is based on the IETF standards which have four layers – Physical Layer (Layer 1 or L1), Media Access Control (MAC) or Transport Layer (Layer 2 or L2), Internet Layer (Layer 3 or L3), and Application Layer. The IETF adapted its internet layers from the Open System Interconnection (OSI) Reference Model which is split into seven layers: Physical, Data link, Network, Transport, Session, Presentation, and Application.

- *Physical Layer (L1):* The Physical Layer, also called L1, is responsible for the actual transmission and reception of raw data over the air interface. It manages functions such as modulation, coding, and signal processing. The physical layer is crucial for establishing and maintaining the radio connection, ensuring that data packets are correctly transmitted over various frequencies. It translates higher-level data into signals that can be sent through the radio medium and handles tasks like error detection and correction. L1 interfaces with Medium Access Layer (L2).
- *Media Access Control Layer (L2):* MAC layer, also termed "Transport Layer," establishes, maintains, and controls connections to transport/exchange information with L1 and Layer 3, which is also called "Logical Layer." MAC performs multiplexing/duplexing of data units and transports to logical channel (L3) and physical layer (L1). L2 performs scheduling, error correction, priority handling, etc.
- *Internet Layer/ Logical Layer (L3):* Data and signaling messages are carried on logical channels. It specifies data packet structure, address, and routing specifications. The internet layer, often referring to the IP layer in LTE, manages the transmission of IP packets, facilitating the connectivity to the internet and enabling services like voice over LTE (VoLTE) and other data services. This layer is crucial for data routing and addressing.
- *Application Layer:* This layer enables end-user services and applications to communicate over the LTE network. It supports interfaces and protocols that applications use to communicate data and manage network services. The application layer is focused on high-level functions that manage user data, with protocols interfacing the L3 using protocols HTTP, FTP, SMTP, and DNS, among others.

7.1.5 Functional Description and Information Flows

As described earlier, protocols specify and manage information exchange which includes control of interfaces, called "Control Plane," and transmission of user data packet, called "User Plane."

7.1.5.1 Control Plane

The control plane consists of protocols for control and support of the user plane functions:

- Controlling the LTE/4G network access connections, such as attaching to and detaching from eNodeB.
- Controlling the attributes of an established network access connection, such as activation of an IP address.
- Controlling the routing path of an established network connection to support user mobility; and
- Controlling the assignment of network resources to meet changing user demands.

The interfaces explained in Figure 7.3 are used among UE, eNodeB, and MME. The air interfaces between UE and eNodeB is called UU (18) and between eNodeB and MME is S1-MME. S1-MME is also termed as S1-C or S1-Contorl.

Non-Access Stratum (NAS): The NAS protocol supports mobility management functionality, user plane bearer activation, and session management, providing security for the communication between the UE, modification, and deactivation. It is also responsible for ciphering and integrity protection of NAS.

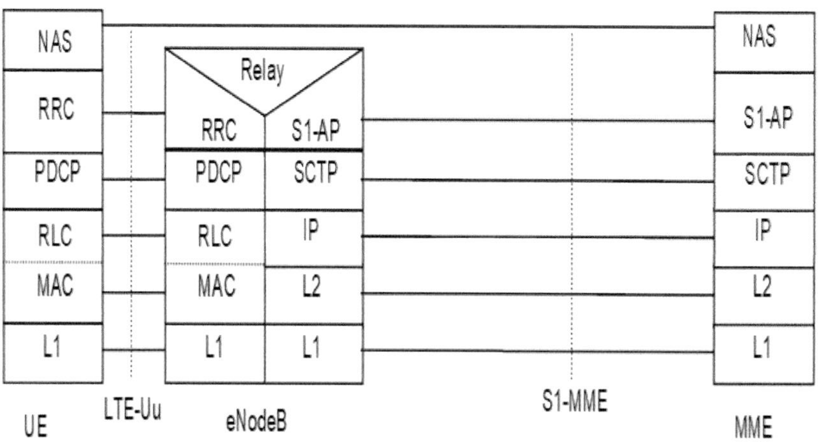

FIGURE 7.3
LTE/4G Control Plane Protocols and Interfaces.

Radio Resource Control (RRC): The RRC is responsible for establishing, maintaining, and releasing connections, as well as handling various functions including security and configuration of radio bearer services.

Packet Data Convergence Protocol (PDCP): The PDCP is responsible for key functions, including the handling of data packets for user plane traffic, header compression to reduce the amount of overhead data transmitted (usually 20 bytes), and providing encryption and integrity protection for the data being sent over the network. PDCP helps improve the efficiency of LTE communications.

Radio Link Control (RLC): The RLC is responsible for the segmentation and reassembly of the packets, error correction through automatic repeat requests (ARQ), and ensures the reliable delivery of data. RLC divides the data received from higher layers into smaller units called Protocol Data Units (PDUs) during the transmission and reassembles them at the receiving end. It operates in three modes: transparent mode, unacknowledged mode, and acknowledged mode.

Media Access Control (MAC): The MAC is responsible for managing how data packets are transmitted over the communication channel. The MAC layer, which is part of transport layer (L2), is responsible for handling tasks such as scheduling, multiplexing, and error correction to ensure that multiple users can access the same physical resources without interference.

S1 Application Protocol (S1-AP): The S1-AP is a signaling protocol to establish and maintain the connection between the UE and the core network. It manages the signaling for establishing, modifying, and releasing user sessions (bearer channel), along with managing resources and mobility procedures.

Stream Control Transmission Protocol (SCTP): SCTP is the transport layer protocol used in the S1-MME interface to provide secure and efficient transmission of signaling messages between the eNodeB and the MME. It provides reliable, connection-oriented data transport and support for message segmentation, reassembly, and error detection. SCTP supports multistreaming and multihoming, enhancing network robustness and efficiency. SCTP ensures the secure and efficient transmission of signaling messages between the eNodeB and the MME.

Internet Protocol (IP): The IP in LTE transmits data as packets utilizing either IPv4 or IPv6 protocols for transmission of data, voice, video, and messaging services.

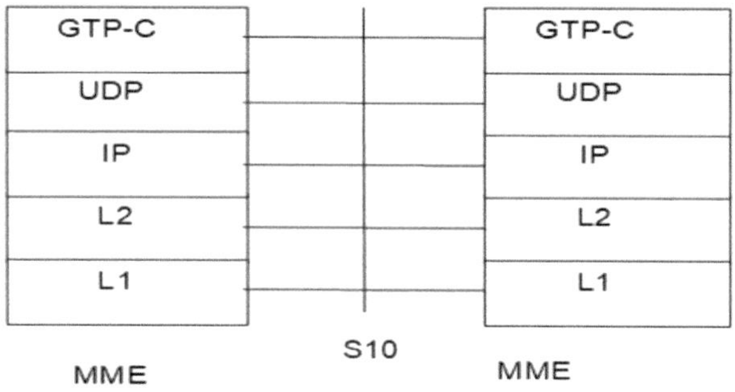

FIGURE 7.4
LTE/4G Control Plane for S10 Interface.

7.1.5.2 MME – MME

In LTE/4G network, S10 interface is used for communication between one MME to the other which is desired when NPSBN MME needs to exchange information with other service providers' network, e.g., for roaming services (see Figure 7.4).

7.1.5.3 MME – HSS

Diameter: This protocol supports transferring of subscription and authentication data for authenticating/authorizing user access to the evolved system between MME and HSS (S6a). Diameter protocol is described in RFC3588 (19).

SCTP: This protocol transfers signaling messages. SCTP protocol is defined in RFC2960 (20).

Transmission Control Protocol (TCP): This protocol transfers signaling messages. TCP protocol is defined in RFC761 (21).

See Figure 7.5.

7.1.5.4 MME – SGW

General Packet Radio Service (GPRS) Tunnelling Protocol – Control Plane (GTP-C): This protocol tunnels signaling messages between MME and SGW (S11).

User Datagram Protocol (UDP): This protocol transfers signaling messages. UDP is defined in RFC768 (22).

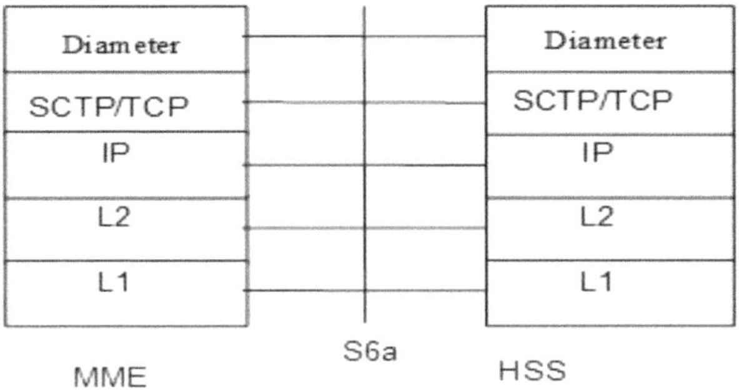

FIGURE 7.5
Control Plane between MME and HSS.

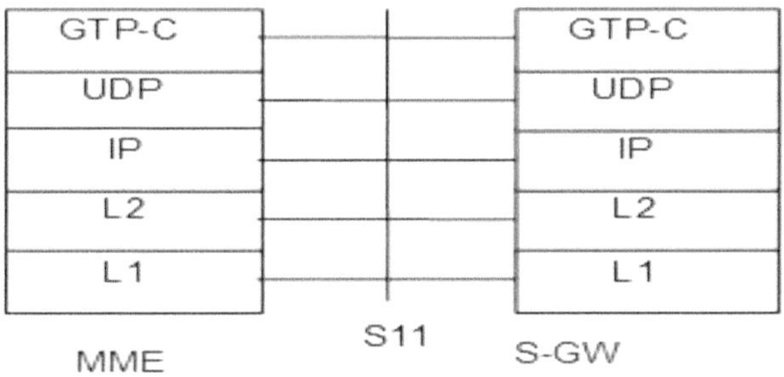

FIGURE 7.6
Control Plane for S11 Interface between MME and SGW.

See Figure 7.6.

7.1.5.5 SGW – PGW

GTP-C: This protocol tunnels signaling messages between SGW and PGW (S5 or S8).

UDP: This protocol user data packets between SGW and PGW. The S5 interface provides user plane tunneling and tunnel management between the SGW and PGW within the same Public Land Mobile Network (PLMN). The S8 interface is used for roaming services and provides user and control plane connectivity between the SGW in the Visited PLMN (VPLMN) and the PDN-GW in the Home PLMN (HPLMN).

See Figure 7.7.

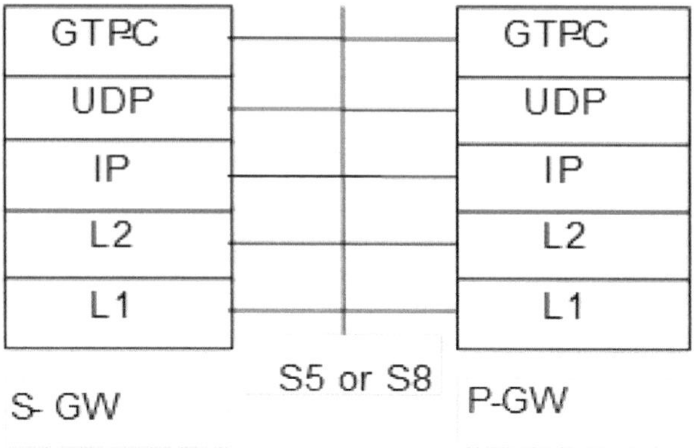

FIGURE 7.7
Control Plane for S5 and S8 Interfaces.

7.1.6 User Plane

The user plane protocols in LTE are basically for packet data transmission between different nodes.

7.1.6.1 UE – PGW User Plane with E-UTRAN

The following paragraphs describe interfaces between UE-eNB, eNB-SGW, and SGW-PGW (see Figure 7.8).

GPRS Tunnelling Protocol – User Plane (GTP-U): The GTP-U encapsulates all end user IP packets and tunnels user data between eNodeB and the SGW as well as between the SGW and the PGW in the LTE/4G network. GTP-U allows for efficient handling of data traffic, providing functionalities such as encapsulation of packets, session management, and optimization of data flow.

User Data Protocol/IP (UDP/IP): UDP/IP protocol is one of the core protocols of the Internet Protocol (IP) suite. It is a communication protocol used across the internet for time-sensitive transmissions. The UDP is connectionless and does not guarantee delivery order or error checking. However, UDP speeds up transmissions because it enables data transfer before the receiving party provides an agreement. As a result, UDP is beneficial in time-sensitive communications. UDP is defined in RFC 786 (23).

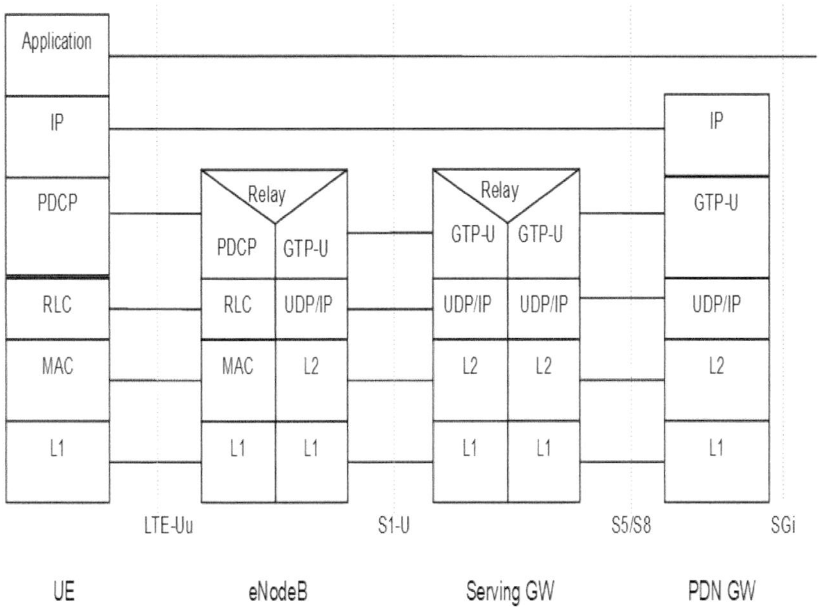

FIGURE 7.8
LTE/4G User Plane Protocols.

Transmission Control Protocol (TCP): TCP is a communication protocol that ensures reliable data transmission between devices over a network. It establishes a connection before data exchange, breaking messages into packets and ensuring they arrive intact and in the correct order. TCP uses mechanisms like error checkin, retransmit lost packets, and acknowledgments to guarantee data integrit.

7.1.6.2 eNodeB – SGW

This protocol stack describes the protocol used in data exchange between eNodeB and SGW over the S1 U interface. The user data is sent in a secure tunnel during the data packet transmission, ensuring data security (see Figure 7.9).

SGW and PGW

The S8 interface is vital for data packet exchange when the user is roaming into a different service provider's network (see Figure 7.10).

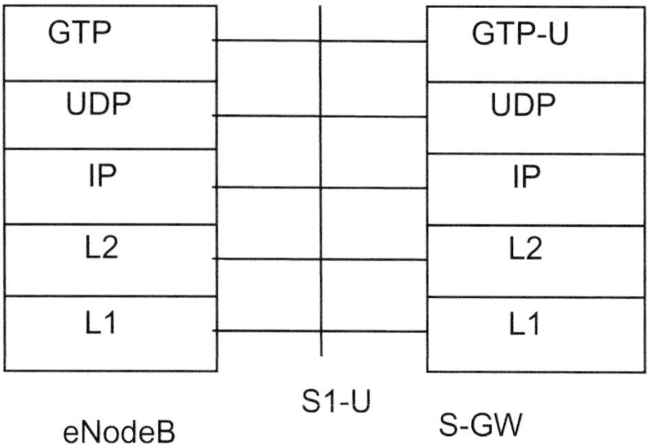

FIGURE 7.9
User Plane for eNodeB and SGW.

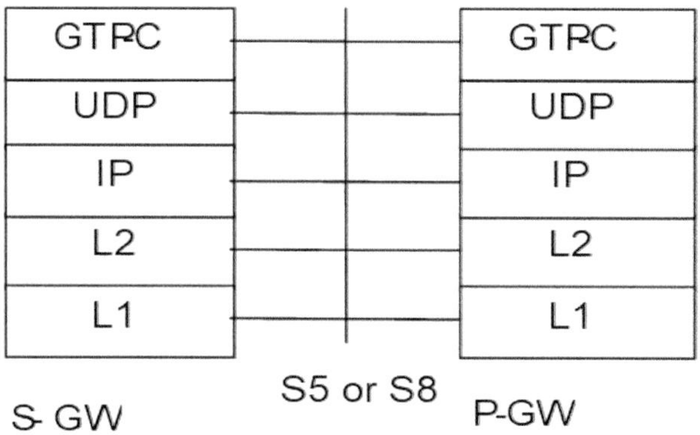

FIGURE 7.10
User Plane S5 and S8.

7.1.7 Initial Attachment Process

This section briefly outlines the protocols and message types used in the initial UE attachment process with eNB and LTE EPC core which are crucial for establishing a user session and ensuring proper connectivity in mobile networks. Specific focus is on the steps and signaling involved for devices initiating a connection with the Evolved Universal Terrestrial Radio Access Network (E-UTRAN) (24) as described in Figure 7.11.

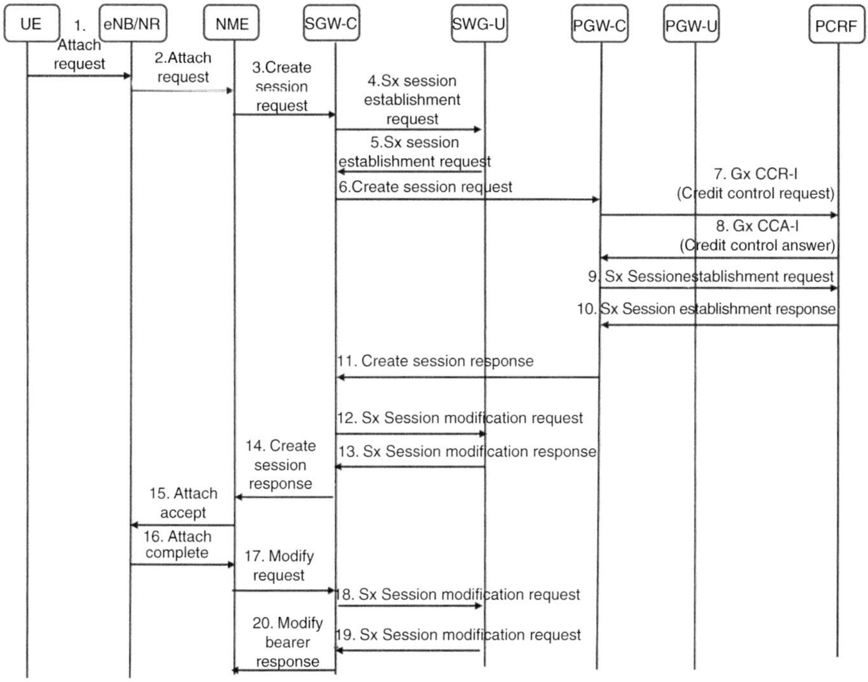

FIGURE 7.11
NPSBN LTE/4G UE Attachment Process.

7.1.8 Data Packet Transmission Procedure

After the initial attachment process described in the previous section, based on NPSBN service APN, desired SGW and PGW are selected. In general, for LTE networks, a secure data packet tunnel is created between PGW and eNB/gNB; however, in NPSBN, a secure encrypted tunnel is created between eNB/gNB and Security Gateway (SGW) in NPSBN.

7.1.9 Deployment and Geographical Locations of NPSBN EPC Core Elements

Consistent with the vision and requirements laid out in the Middle Class Tax Relief and Job Creation Act of 2012, NPSBN is built using cutting-edge technology (e.g., 4G/5G) to provide secure, prioritized, and preemptive wireless broadband communications to millions of first responders at local,

TABLE 7.4
NPSBN Core Centers

NPSBN Geographic Regions	Primary NPSBN Core Center	Secondary NPSBN Core Center
Northeast	Broadway	Alpharetta
Southeast	Alpharetta	Broadway
North Central	Akron	Allen
South Central	Allen	Akron
Northwest	Bothell	Concord
Southwest	Concord	Bothell

state, tribal, and federal level across all states, territories, and the District of Columbia. In addition, in view of the critical importance of the NPSBN for first responders to save life and properties, it is most desirable that communications must be available locally encountering the least latency and with QoS required for NPSBN.

In mainland United States, three regions serve first responders to meet latency requirements. Here is the list of regions.:

East Region
Central Region
West Region

Each region has required geographical redundancy to meet the local availability, latency, and QoS.

For location and location-level redundancy in a primary/secondary/tertiary model, the closest location will generally be chosen based on the APN DNS mapping (see Table 7.4).

Figure 7.12 provides the locations of the EPC elements. It should be noted that the states of Alaska, Hawaii, and Caribbean territories are all served by EPC cores deployed in the U.S. Mainland.

7.1.10 NPSBN Coverage Areas

FNA has an internet portal accessible to public for visualizing the NPSBN coverage in any part of the U.S. mainland (25). It should be noted that the NPSBN coverage area map is updated regularly to reflect the most current coverage status.

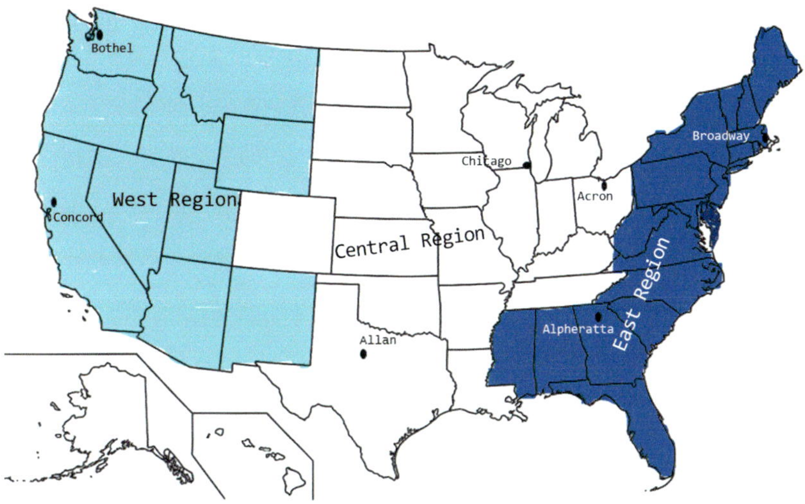

FIGURE 7.12
NPSBN EPC Elements Deployment Locations.

7.2 5G CORE NETWORK TECHNOLOGY EVOLUTION

5G technology is the next generation of mobile technology, which is still evolving and maturing. The timeline of 5G network development includes several key milestones:

2012: The ITU began discussions on what would become 5G technology.
2015: The ITU officially defined the requirements for 5G in the IMT-2020 report.
2016: The first 5G standardization efforts started, led by the 3GPP.
2017: 3GPP completed the first phase of 5G specifications (Release 15).
2019: Major carriers in United States such as Verizon, AT&T, and T-Mobile launched commercial 5G services in select cities.
2020: The ITU officially approved the standards for 5G networks.
2021: Global 5G subscriptions surpassed five hundred million, with significant infrastructure growth.
2022: Continued expansion of 5G networks, with enhancements in speed and coverage.

The timeline highlights the rapid development and deployment of 5G technology, which has potential benefits, especially in the following areas:

High-speed packet delivery
Low latency
Higher bandwidth
Edge slicing
Increased coverage
More connected devices

7.3 5G CAPABILITIES FOR NPSBN

Enhanced mobile broadband (eMBB): 5G enhances multimedia experiences with higher bandwidth allocation and throughput.

Ultrareliable, low-latency communication (uRLLC): 5G supports requirements for mission-critical and latency-intolerant applications with sub-millisecond latency.

Massive machine-type communications (mMTC): 5G empowers cost-efficient and robust connection for large numbers of devices without network overloading.

The new business opportunities made possible by 5G are vast, which would span across industries. Some of these opportunities include:

- *Telecommunications, media, and entertainment:* Network maintenance and optimization, self-learning security, neural nets, and personalized content
- *Automotive:* Predictive maintenance, autonomous ride sharing, and driver assistants
- *Energy:* Predictive infrastructure maintenance, grid operations and storage, and smart metering
- *Financial services:* Trading, personal finance planning, fraud detection, and customer operations
- *Healthcare:* Enhanced diagnostic and image analysis, early pandemic detection

- *Manufacturing:* Process correction, optimization, and on-demand production
- *Retail and consumer:* Personalized production, demand prediction and analysis, and delivery

7.3.1 5G Network Architecture and Subsystems

Similar to LTE/4G network architecture, 5G has a robust architecture to support the next generation of mobile communication with emphasis on high speed, low latency, and significantly increased coverage. The standards and recommendations for 5G requirements, architecture, and services are generally published by 3GPP through joint work with many national and international standards bodies (e.g., ISO, ITU, ETSI, IEEE) and consortiums of national or regional standards organizations and private companies which are actively engaged in developing technical specifications for mobile networks including 4G/LTE, 5G, and future 6G networks.

The 3GPP 5G architecture includes several key components, such as 5G capable UE, gNodeB (the 5G base station), the network transport to various destinations, the 5G core (5GC and content/applications to support end-to-end connectivity and desired outcome. The architecture is designed to accommodate both native 5G services and ensure the interoperability with existing technologies through a service-based architecture (SBA).

Figure 7.13 illustrates a high-level 5G architecture comprising essential subsystems. Only 5G capable devices would be able to attach and

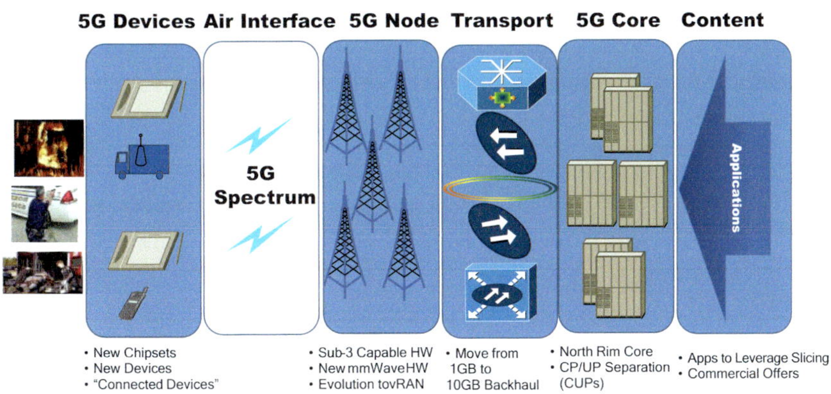

FIGURE 7.13
5G Network Framework for the Layered Architecture.

communicate with 5G RAN and access 5G core to provide the desired outcome of the applications.

7.3.2 FN NPSBN 5G Core Network

5G core network incorporates SBA, which allows for greater flexibility and scalability compared to previous generations as a connectivity pipe for service delivery. This architecture enables network functions (NFs) to communicate with one another through defined APIs, enhancing interoperability. Interfaces in the 5G CN are denoted by adding "N" followed by the lowercase name of the specific NF (e.g., AMF interface is named Namf). Using the named interface, any other NF can be able to communicate and exchange the data packet with the specific NF. In addition, 5G core supports a wide variety of applications including eMBB, massive machine-type communications, and uRLLC. It also incorporates control plane and user plane separation technology services or customer requirements, and cloud-based deployment with the ability of to use centralized or distributed elements for better performance and management.

Figure 7.14 illustrates 5G network core incorporating SBA.

5G SBA core network is geared towards separation of the service functionalities from the network logic enabling a modular design and implementation to suit mobile service providers, their unique business models, and clients. As illustrated in the figure, NFs are generally implemented virtually in cloud as Virtualized Network Functions (VNF) and uses API calls to invoke specific service as desired by live users or by Machine to Machine (M2M) in Internet of Things (IoT) applications. 5G Service Functions (SF) are likely to be expanded as new types of service in public safety, manufacturing, M2M, and self-driving cars are added. In 5G networks, the MME is replaced by the Access and Mobility Management Function (AMF), which provides broader functionalities including:

- The AMF is responsible for managing the mobility of subscribers within the network, authenticating subscribers, and ensuring the security and QoS for subscribers.
- The AMF communicates with several other network elements within the 5G architecture, including the user plane function, Session Management Function (SMF), and Authentication Service Functions (AUS).
- The AMF is responsible for managing the handover of subscribers between different base stations and ensuring that the subscriber's connectivity is maintained during the transition.

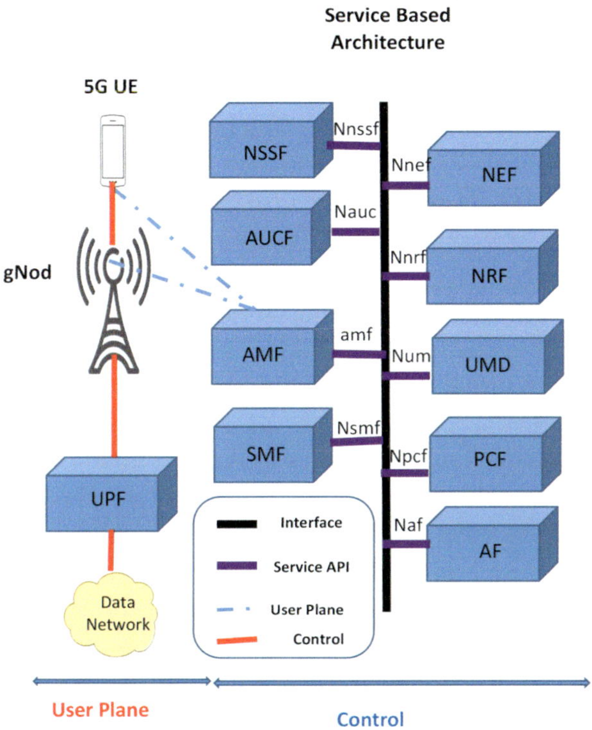

FIGURE 7.14
5G Core Network – SBA.

- The AMF is also responsible for managing the routing of data packets to the appropriate UPF based on the subscriber's location and QoS requirements.

UPF is a key function responsible for handling user data traffic, which includes the routing and forwarding of data packets between the UE and the DN. The main function of the UPF includes:

Data Forwarding: UPF forwards user data packets to and from the UE and the external DNs, such as the internet or private networks.

Traffic Management: It manages QoS for user data flows, ensuring that diverse types of traffic receive appropriate bandwidth and latency characteristics.

Packet Inspection: UPF can perform deep packet inspection to analyze the data packets for various purposes, including billing, security, and traffic management.

- *Policy Enforcement:* UPF enforces policies defined by the network operator, such as traffic shaping and prioritization based on user subscriptions and service-level agreements.
- *Interfacing with Other Functions:* UPF interacts with other NFs, such as the SMF for session establishment and management, and the AMF for user authentication and mobility management.
- *Support for Network Slicing:* UPF plays a crucial role in network slicing, allowing multiple virtual networks to operate on a single physical infrastructure, each tailored to specific service requirements.

3GPP SBA architecture also has additional service functions nodes including:

Network Slice Selection Function (NSSF) enables multiple virtual networks to be created on a single physical 5G infrastructure. Each slice can be customized to meet the specific requirements of different services or applications, such as low latency for gaming, high bandwidth for video streaming, or massive connectivity for IoT devices. NSSF enables operators to efficiently manage resources and provide tailored services to various users and industries.

Network Exposure Function (NEF) serves as a pivot node that enables the exposure of functionalities such as network information, user sessions, and policies to external applications, facilitating the development of new services and applications that can leverage 5G capabilities.

Network Repository Function (NRF) of 5G SBA manages the registration of NFs and facilitates the discovery of required services between these functions. It is a central registry where various NFs can register their capabilities and discover other functions they might need to communicate with, ensuring seamless operation within a 5G network. NRF enhances network flexibility and scalability by enabling dynamic interactions among different network services.

Unified Data Management (UDM) in 5G is a framework that centralizes the management of user data and service information across the network. It integrates various data types, such as user, subscription, and application data, allowing for more efficient data handling and policy management in a 5G environment. UDM helps enable better service provisioning, QoS, and overall network automation.

Policy Control Function (PCF) is a network function that manages policy rules and controls resource allocation for user sessions. It ensures appropriate QoS, applies usage policies, and coordinates with other NFs to optimize network performance and user experience. The PCF interacts with various components, such as the NEF and the UDM, to enforce policies based on the network conditions and user demands.

Application Functions (AF) provides application-layer services to users and applications. They facilitate the connection between applications (such as web services, streaming, and IoT applications) and the 5G core network by managing user sessions, enabling service provisioning, and ensuring QoS. AF also plays a crucial role in slicing, allowing different applications to use tailored network resources and performance characteristics

7.3.3 Control Plane and User Plane Separation

One of the specific features incorporated into 5G architecture is the CUPS. CUPS in 5G refers to the architecture that separates the control functions from the data transfer functions within the network.

Control Plane: This part of the CUPS handles signaling and management tasks, including the setup, maintenance, and teardown of connections. It takes care of the signaling related to the UE and network operation.

User Plane: This part of the CUPS is responsible for the actual data transfer between the UE and the network. It carries user data traffic and is optimized for performance and low latency.

CUPS allows for scalability and flexibility, enabling operators to optimize resource allocation, improve overall network efficiency, and support different use cases in 5G.

Figure 7.15 illustrates the 5G CUPS architecture showing control plane and user plane functions.

The architecture for CUPS of SGW, PGW, and TDF is based on the following concepts:

- Interworking with networks not applying CUPS is possible (i.e., in roaming scenarios)
- Split network entities can interwork with network entities that are within the same network
- Split network entities have no requirement to update UE and RAN

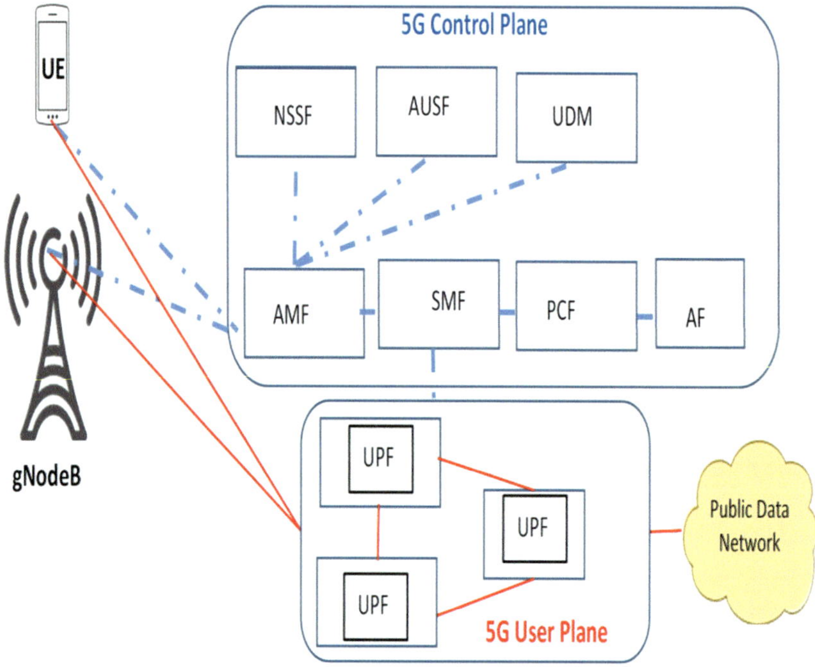

FIGURE 7.15
5G CUPS – Logical Architecture.

- The SGW/PGW selection function of the MME/ePDG/TWAN described in TS 23.401 (26) and TS 23.402 (27) is used for the selection of the respective CP function.
- The configuration-based mechanism (in PGW or PCRF) described in TS 23.203 (28) is used for the selection of the CP function of the TDF.
- A CP function can interface with one or more UP functions (e.g., to enable independent scalability of CP functions and UP functions).

7.3.4 5G Deployment Categorization

The previous wireless network evolution (2G, 3G, and 4G) required both RAN and core network of the same generation to be deployed together. However, in 5G network, there is flexibility to integrate elements of different generations in different configurations. 3GPP (TS 23.401) (29) has specified several 5G deployment options that can be categorized as either standalone (SA) or non-standalone (NSA) architectures.

7.3.4.1 5G SA Deployment

SA deployment in 5G refers to a network architecture that operates independently of any previous generation networks like 2G, 3G/UMTS, 4G/LTE. In SA, the UE connects to the 5G RAT and attaches to the 5G core network, allowing operators to take full advantage of 5G's capabilities, such as ultra-low latency and high bandwidth. This deployment supports the complete range of 5G services without reliance on or interference from older technologies. 5G SA architecture also introduces Data Network Name (DNN), a new way of requesting connection to the core for IP address assignment. DNN is synonymous with APN in 4G/LTE. Figure 7.16 shows the simple framework of 5G SA deployment.

Compared to the current LTE/NSA core deployments, the control and user plane locations for the SA solution will be further distributed within a region. An example of such a deployment could consist of four control plane locations and multiple user plane geo-distributed locations in the region, as opposed to the limited number of locations in the current LTE/NSA deployment.

7.3.4.2 5G NSA Deployment

The NSA deployment integrates 5G RAT with existing 4G/LTE infrastructure. In this configuration, 5G services rely on the 4G core network for managing user data and signaling, enabling mobile operators to provide 5G services more quickly while leveraging their existing 4G infrastructure. This approach is also generally termed Dual Connectivity, specifically called E-UTRA New Radio Dual Connectivity (ENDC), which is the primary option used in the NSA architecture. ENDC feature in 5G technology allows a user

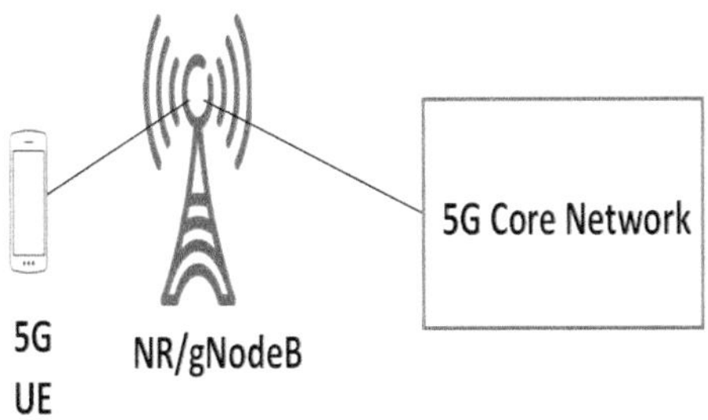

FIGURE 7.16
5G SA Deployment.

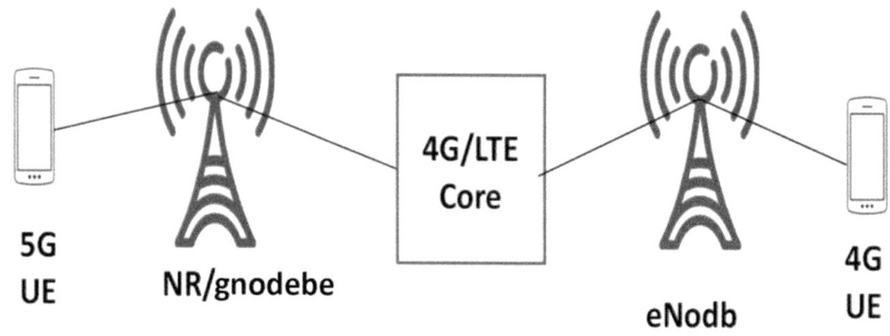

FIGURE 7.17
NPBN 5G NSA Deployment – Logical Architecture.

device to connect to both LTE (4G) and 5G NR (New Radio) networks simultaneously. This enhances the user experience by providing greater bandwidth and lower latency. Mobility is achieved between LTE and SA via Inter-Radio Access Technology (IRAT) mobility procedures between the EPC and 5GC. This allows for enhanced speed and lower latency without the need for a complete overhaul of the existing network.

There are several variations of NSA deployment that service providers opt to meet the existing network and business requirements. Figure 7.17 illustrates a deployment of 5G NSA where 5G users (with 5G-capable UE) and 4G users (with LTE UE) can access NR/gNodeB and eNodeB, respectively. However, both users access the LTE core for their connectivity and services.

NPSBN 5G deployment strategy is to fully deploy SA in near future to avail full advantage of the 5G technology. Currently, NPSBN uses NSA deployment having B14 spectrum which is allocated by Federal Communications Commission (FCC) to FirstNet. FNA is working with its private partner, AT&T, to migrate from NSA to SA using n14 spectrum capable of 5G broadcast.

7.3.5 5G Core Network Geographical Deployment

Currently, NPSBN geographical location and 5GNC element redundancy strategy is to distribute CP and UP instances over multiple locations in each region to ensure coverage uniformity in both urban and rural areas are met to manage natural and man-made disasters to confirm NPSBN's 24×7 availability.

7.3.6 NPSBN 5G Coverage Area

FN/AT&T is diligently progressing to deploy 5GSA which is expected to be available by 2026. FNA has an internet portal accessible to public for visualizing the NPSB coverage in any part of the U.S. Mainland (25).

7.4 NPSBN 6G NETWORK

As stated earlier, 6G is the future generation of wireless communication network evolution (30). 6G will pave the way for enhanced public safety services, communications with ultra-low latency and high reliability, high data packet size transmission, and AI for performance and reporting.

7.4.1 6G Platform Features – A Key Enabler of Complex Communications and Services

6G network will be a key enabler platform to develop complex emergency management services utilizing AI and Machine Learning (ML), adding new devices and sensors, and many more. In addition, it will handle network management, enable massive device connectivity (massive IoT), and provide robust security measures. A few key features that will be enabled by 6G is listed here:

- Flexible multiband utilization
- Super flexible PHY and MAC
- Flexible heterogeneous networks
- Integrated sensing and communications
- Green communications
- Secure network and cybersecurity
- Use of AI and ML

Figure 7.18 provides a futuristic 6G communications incorporating many sensing devices, devices, and integration.

Figure 7.18 provides a strategic long-term and integrated service profile capabilities planning for 6G. The outer layer of the semicircle describes several types of sensing devices operating on ultraflexible heterogeneous networks including terrestrial, aerial, satellite, and undersea networks. The innermost

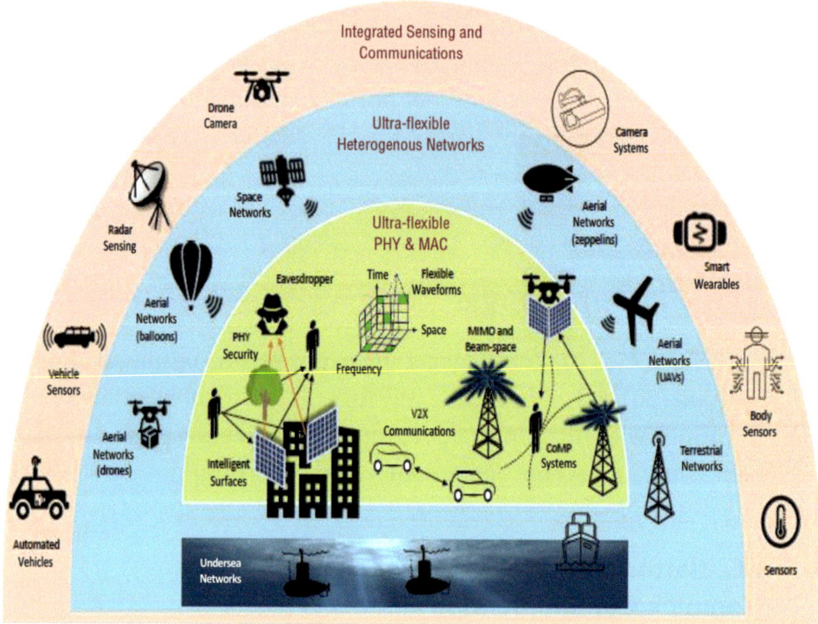

FIGURE 7.18
6G Vision and Strategy for Heterogeneous Communication Networks.

layer comprises PHY and MAC layers to support computing environment for various public service applications.

7.4.1.1 6G Timeline

There is no officially announced launch time for 6G by any service providers yet. The standards communities and industries are still working towards finalizing the 6G vision, requirements, architecture, spectrum, etc. The consensus seems to be that discussions on the exact technical specifications will start in 2025, leading to the first 6G specification in 3GPP Release 21 by 2028. Commercial deployments may begin around 2030.

7.5 NPBN CORE – KEY POINTS

The NPSBN core network, also known as the EPC, is a critical component of LTE architecture that enables high-speed data transmission and supports numerous services. The NPSBN core combined with IMS is the heart of the broadband network and supports voice, video, and data services. Initially, NPSBN deployed the LTE/4G core which interfaces with PDN, Private Service Telephone Network (PSTN), internet, and other service provider's networks. FN subscribers connect NPSBN core using their FN mobile devices over 4G RAN to avail desired NPBN services. The LTE RAN is responsible for radio communications with the UE, including radio resource management, handovers, and managing radio bearers. The main NPBN core elements are:

MME: Manages mobility-related functions such as tracking, paging, and handovers. It also authenticates users and initiates bearer activations.

SGW: Routes and forwards user data packets. It acts as the mobility anchor during inter-eNodeB handovers.

PGW: It acts as the gateway between the LTE network and external PDNs such as the internet. It performs IP address allocation and manages QoS.

HSS: This is the subscriber database that stores user subscription information, including authentication information, user profiles, and service subscription data.

PCRF: This is responsible for policy control and charging functions. It manages the allocation of network resources based on policy rules and determines charging parameters for the user's data usage.

Recently, 5G technology is getting actively designed and deployed. It has potential benefits especially in the following areas:

High-speed packet delivery
Low latency
Higher bandwidth
Edge slicing
Increased coverage
More connected devices

5G is a transformative technology focused on SBA as compared to 4G which is connection-based design. In addition, 5G takes advantage of AI and ML to support IoT and MTM services. NPSBN already has non-SA (NSA) 5G network deployed. It is expected that NPSBN will launch SA 5G-based NPSBN by 2026.

The 3GPP and other national and international standards organizations are actively discussing 6G which is the next generation of evolution in network technology to support capabilities like Flexible multiband utilization and heterogeneous networks, enhanced AI and ML-based new services, secure network and cyber security, energy efficient and ultrafast network. 6G technology is expected to be ready to be deployed by 2030.

8

NPSBN Transport Network

National Public Safety Broadband Network (NPSBN) transport network involves the architecture, design, and protocols used to enable efficient data transmission over mobile networks. It primarily utilizes an IP-based architecture, Ethernet, and fiber-optics transport media ensuring high-speed data transfer by minimizing latency and leveraging key technologies such as Multiple Input Multiple Output (MIMO), Orthogonal Frequency Division Multiplexing (OFDM), and various Quality of Service (QoS) mechanisms.

Figure 8.1 depicts high-level NPSBN transport framework interconnecting various NPSBN subsystems – Radio Access Network (RAN), LTE core network, and private/public data network (PDN) (see Figure 8.1).

8.1 NPSBN END-TO-END TRANSPORT ARCHITECTURE

Figure 8.2 provides an end-to-end logical architecture depicting various transport segments.

Here is a brief description of each segment.

1. *Radio Wave:* The segment between FirstNet UE and 4G/5G. RAN is over the air on radio waves, hence no physical wire connection is required.
2. *Front-Haul:* The transport segment between 4G/5G RAN and the aggregation node, called Smart Integrated Access Device (SIAD). A SIAD is a networking equipment designed to provide multiple services, such as internet access, telephony, and video streaming over a single unit. SIADs facilitate the integration of various technologies

DOI: 10.1201/9781003557739-10

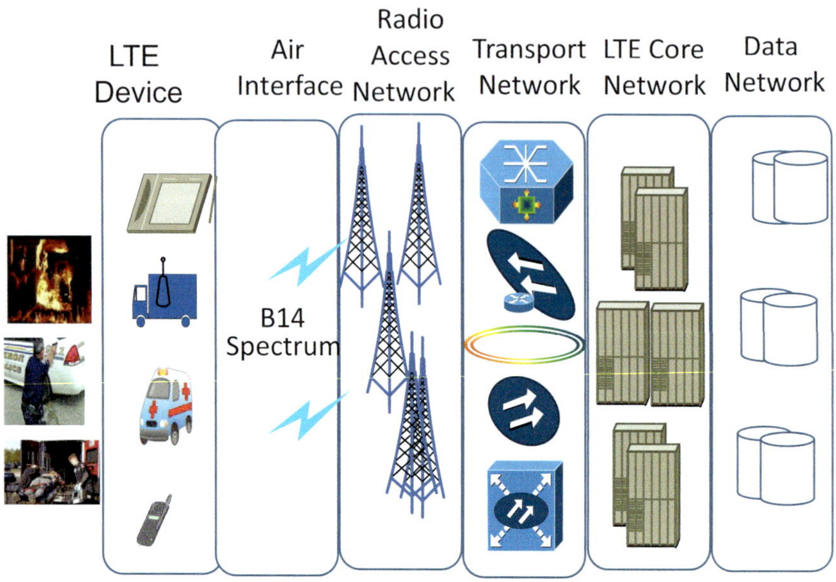

FIGURE 8.1
High-Level Logical Framework of 4G/5G Transport.

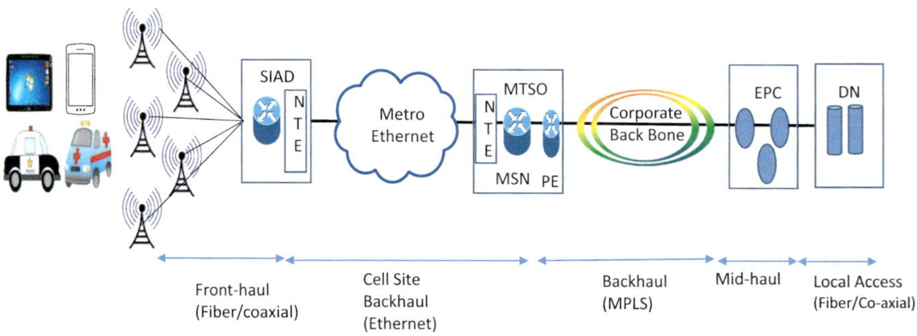

FIGURE 8.2
NPSBN Transport Segments and End-to-End Transport Architecture.

such as broadband, VoLTE, and IPTV, enhancing the efficiency and performance of communication services. SIAD provides advanced management capabilities, enabling service providers to deliver tailored solutions and manage their network resources better. SIAD connects to Network Termination Equipment (NTE). The NTE connects the Cell Site Backhaul (CSBH). NTE includes the hardware and software necessary to manage network connections and ensure data transmission.

3. *CSBH:* The transport segment that spans between SIAD and the Mobile Station Node (MSN) deployed in Mobile Telephone Service Office

(MTSO). In LTE networks, MSN is an important switching node that connects mobile devices to the core network and other service networks, managing calls, data sessions, and providing mobility between various base stations. A pair of MSN and Provide Edges (PEs) are housed in MTSO. MSN facilitates call control and routing of multimedia and voice services. It ensures seamless connectivity and communication between different network elements, managing the signaling and data trafficking to support Voice over LTE (VoLTE) and other services. Essentially, it plays a critical role in the overall management of the mobile network's communication processes. The PE is a router/switch node, which is housed in the MTSO, and directs data packets based on IP addresses and determines the best path for data transmission across networks.

4. *Backhaul:* The major segment of the LTE transport that spans between the MTSO and the Evolved Packet Core (EPC) office. It typically involves high-capacity transmission links that carry the data from UE/RAN/SIAD to the core network. A typical backhaul can be implemented using various technologies such as fiber optics, microwave links, or copper lines, and is crucial for managing the traffic generated by mobile users and ensuring efficient data transfer. NPSBN backhaul, called Common Backbone (CBB), uses Multiprotocol Label Switching (MPLS) technology. MPLS is based on the self-healing optical ring design and routing data from one node to the next based on short path labels rather than long network addresses. This enhances the speed and efficiency of data transmission across networks, supports diverse types of traffic, and facilitates the creation of Virtual Private Networks (VPNs), traffic engineering, and QoS management
5. *Mid-Haul:* A segment that is not generally mentioned or well specified in most of the literature. However, it is a transport segment spanning between backhaul and the LTE EPC centers. Generally, it is point-to-point or multipoint connections over fiber cable connecting backhaul/MPLS to specific EPC node.
6. *SGi Interface Segment in Transport:* The transport segment that connects the EPC and the PDN referred to as the "SGi interface." This interface facilitates the data transfer between the core network and external data networks. The transport segment usually uses fiber-optic or copper cables and generally deploys a VPN technology that creates a secure and encrypted connection over a less secure network, such as the internet. It allows users to send and receive data as if their devices were directly connected to a private network, enhancing privacy and security by masking the user's IP address and encrypting their internet traffic.

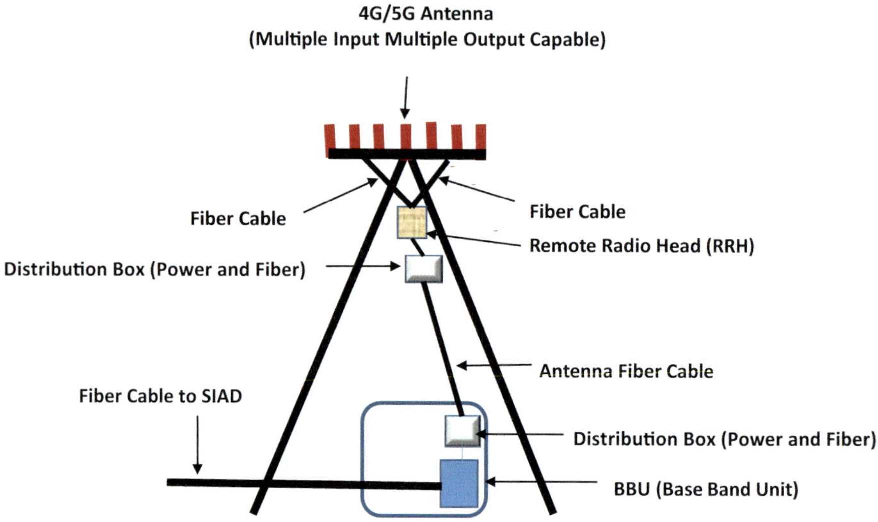

FIGURE 8.3
NPSBN RAN Tower.

8.1.1 RAN Connectivity

4G/5G RAN is a complex system comprising radio signal receiving, broadcasting, and MOMO capacity to receive and broadcast many signals simultaneously, proving network security and QoS. A 4G/5G antenna is a specialized RF design optimized to receive and transmit signals in different frequency bands, e.g., B14 for 4G and n14 for 5G. Figure 8.3 illustrates a simplified view of the connection of the NPSBN RAN components. It should be noted that most of the RANs are shared among various mobile service providers; therefore, the RAN tower is designed to provide a varied range of services to different clients based upon their business requirements. FNA partner AT&T has over a thousand RAN towers (owned and leased) which is shared between NPSBN and AT&T clients.

8.1.2 Metro Ethernet Network

Architecture

Metro Ethernet is a transport network (Layer 2/Data Layer) that provides point-to-point or multipoint connectivity services over a metropolitan area network (MAN). Ethernet network originated as Virtual Local Area Network (VLAN) technology and became a replacement for low-speed WAN technologies. NPSBN CSBH transport is based on Ethernet VLAN as shown in

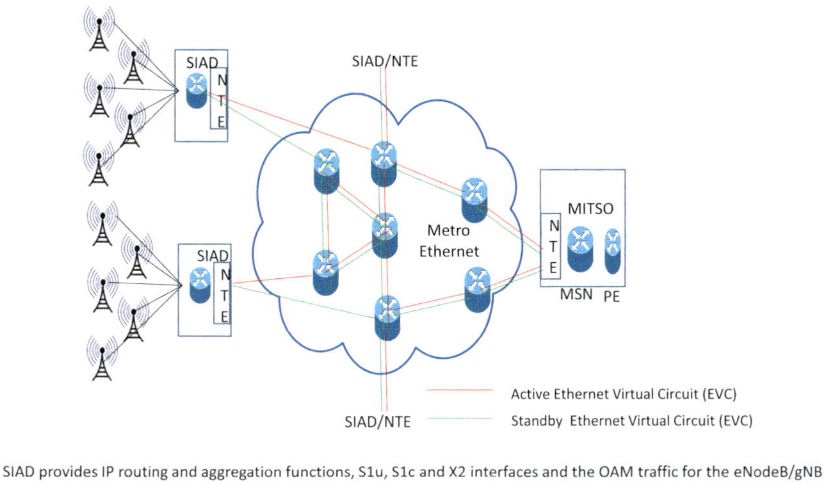

FIGURE 8.4
NPSBN CSBH Ethernet Transport Architecture.

Figure 8.4. Ethernet VLAN connects to several SIADs to connect to MTSO/ MSN to transport data packets.

Ethernet is a reliable, secure, and standards-based network to support NPSBN CSBH transport segments capable of 100 Gbps (giga bit per second) and can be implemented in network cloud deployment. The main features and benefits of the Ethernet network is summarized here:

- *Flexibility:* Supports a wide variety of services and transports. Since it is Layer 2 network, it can be deployed on different Physical Layer, e.g., fiber or copper cables.
- *Reliability:* Ethernet operations, administration, and maintenance (OAM). It performs path discovery, detects, and reports connection failures and measures performance.
- *Cost Effectiveness/Ease of Use:* Networks are less complicated and easier to maintain than WAN networks, which lowers equipment and ownership costs.
- *QoS:* Supports QoS features, such as classification, marking, policing, queuing, and scheduling.
- *Scalability:* Supports speeds from one Mbps to 100 Gbps. Customers can increase bandwidth dynamically, and often without the need to purchase or install new equipment.
- *Protection:* Facilitates alternative switching paths and enables 50-msec protection by using labeled switch-path tunnels.

- *Optimization:* Allows providers to optimize their bandwidth for protected traffic and to allocate less bandwidth for unprotected traffic.

Ethernet VLAN Design

VLAN operates primarily at Layer 2 of the OSI model, which deals with data link protocols. VLAN allows the segmentation of a single physical network into multiple logical networks. However, when VLANs are routed, they can also operate at Layer 3, where IP addressing and routing take place using an L3 subnet. For two devices to communicate across a typical business or home network, they need to have both an IP address, associated with Layer 3 (the IP layer), and a MAC address, associated with Layer 2 (the Ethernet layer).

In NPSBN, connection between Base Band Unit (BBU) and SIAD are over VLAN. Separate VLANs are used between BBU and SIAD for different types of traffic described here:

- VLAN 1 is used for S1 U, S1 C, and X2 traffic.
- VLAN 2 is used for OA&M traffic
- VLAN 3 is used for Precision Time Protocol (PTP) for clock synchronization backup in case GPS fails

Ethernet Virtual Circuit Design

An Ethernet Virtual Circuit (EVC) is a conceptual channel/pipe within the service provider network (i.e., VLAN) and defines a Layer 2 bridging architecture that supports Ethernet services. EVC is defined by the Metro Ethernet Forum (MEF) as an association between two or more user network interfaces that identifies a point-to-point or multipoint-to-multipoint path within the service provider network.

NPSBN CSBH transport architecture utilizes the following three specific EVC to support first responders:

1. Active EVC: Has established connectivity between SIAD and MSN and carries data packet.
2. Standby EVC: A secondary or standby EVC takes over the data packet transmission in case active EVC fails.
3. OAM EVC: Carries information relating to status and performance of the data transmission in EVC.

Since the scope of this book is not to dwell in detail on the design, routing, and protocols of Ethernet networks, interested readers are referred to the Ethernet standards specifications. A few key standards are listed here:

1. Ethernet MEF10.2 and 10.21
2. Institute of Electrical and Electronics Engineers (IEEE) 802.3: Standards for Ethernet networks, covering aspects such as frame format, physical layer specifications, and protocols.
3. IEEE 802.1Q: Standards for VLAN tagging to support traffic separation in Ethernet networks.
4. International Telecommunication Union – Telecommunication Standardization Sector Y Series (ITU-T) Y.1564: A standard used for Ethernet service activation testing.
5. ITU-T G.8031 and G.8032: Standards for Ethernet service protection, including linear and ring protection.
6. IEEE 802.3ah: Standards for Ethernet in the First Mile (EFM) which defines specifics for last-mile broadband access.

These standards help ensure interoperability and service quality in Metro Ethernet deployments. For detailed compliance and standard updates, checking with the IEEE and ITU-T organizations is recommended.

8.1.3 NPSBN Backhaul

AT&T shares its IP/MPLS backhaul transport network, namely Core Backbone (CBB), among all its subscribers including the first responders who use NPSBN network. It is an added advantage to NPSBN users to readily utilize AT&T's vast transport network covering the entire Continental Unites States. Going onward, NPSBN Backhaul will refer to the shared AT&T CBB network. Although the NPSBN backhaul is shared network, first responders have specific QoS criteria, which is described later.

NPSBN backhaul Layer 1 (physical media) is basically built on self-healing fiber-optic ring architecture using Dense Wave Division Multiplexing (DWDM). AT&T is a pioneer in deploying Fiber to the Curve (FTTC) and Fiber to the Premise (FTTP) networks which provide high-speed internet access. In FTTC the fiber-optic cables run to a point near customers, typically a curb, and then the final connection is made via copper lines. FTTP, on the contrary, refers to direct fiber-optic connections going to home and apartment buildings in urban areas for faster internet speeds. Both services aim to deliver reliable and speedy internet connections using fiber optics as a physical mode.

International Telecommunication Union (ITU) has termed DWDM network as Optical Transport Networks (OTN) (31, 32). OTN topology consists of the Optical Network Element (ONE).

8.1.3.1 ONEs

8.1.3.1.1 Optical Transponders

An optical transponder in OTN is a device that converts electrical signals originating from the client service into optical signals and vice versa and consists of a transmitter and a responder, which is like a transceiver that includes a transmitter and a receiver. It allows multiple data channels to be transmitted simultaneously over a single optical fiber by utilizing different wavelength channels, thus increasing the capacity of the fiber-optic network. Optical transponders are essential for enabling long-distance communication in telecommunication networks utilizing fiber-optics transport – WDM/DWDM (see Figure 8.5).

The optical transponder usually combines with other OTNs to automatically receive, amplify, and then retransmit a signal on a different wavelength without changing the data/signal content.

Here are the key functions of a transponder:

1. Signal Conversion: The primary function of a transponder is to convert electrical signals into optical signals for transmission over fiber-optic

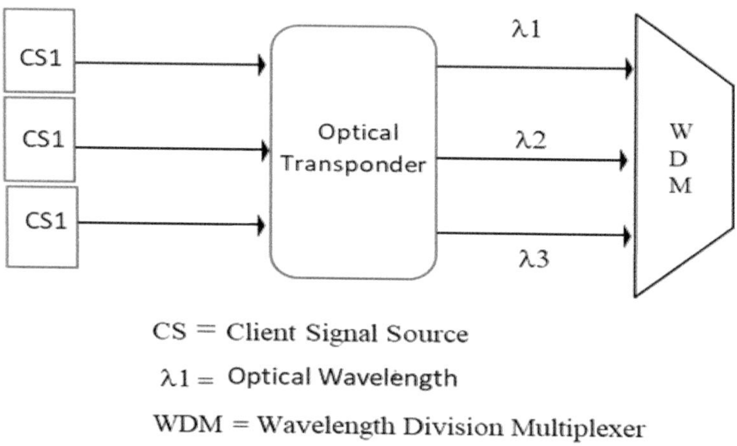

FIGURE 8.5
OTN Transponder Design.

cables. Upon reaching the destination, it converts the optical signals back into electrical signals.

2. Wavelength Assignment: Transponders can assign different wavelengths to signals. This process, known as wavelength division multiplexing (WDM), allows multiple signals to be transmitted simultaneously over a single fiber, significantly increasing the network's capacity.
3. Signal Regeneration: Over long distances, optical signals can degrade. Transponders regenerate these signals to ensure they remain strong and clear, thus maintaining the integrity of the data being transmitted.
4. Error Correction: Advanced transponders are equipped with error correction mechanisms to detect and correct errors in the data stream, enhancing the reliability of the transmission.

Optical transponders are versatile for converting several types of signals, including multimode to single mode, dual fiber to single fiber, and one wavelength to another wavelength. During the process, the signal is also reamplified, monitored, and cleaned up but will not be modified to be different from its original bit rate in order to extend the transmission distance. NPSBN OTN uses single fiber-optical transceivers in which the transmission and reception of data is conducted on a single fiber. Single-mode fiber-optic cable is designed to carry light directly down the fiber with minimal interference. It has a small core diameter (typically around 8–10 micrometers) that allows only one mode of light to propagate, making it ideal for long-distance communication as it minimizes modal dispersion. This results in higher bandwidth and longer transmission distances compared to multimode fiber. Dual fiber-optical transceivers use the same wavelength on two fiber strands. It has two distinct channels or ports for transmission and reception of data

8.1.3.1.1.1 Optical Mux/Demux Figure 8.6 illustrates the optical multiplexer, that is, the operation of a WDM. In this figure, the direction of transmission is the same for all wavelengths. Systems of this type are called unidirection WDM systems. A full duplex operation would require another system for the return channels. WDM systems in which both directions of transmission are carried by the same fiber are called bidirectional WDM systems.

8.1.3.1.1.2 Optical Fiber Amplifier An optical fiber amplifier is a ONE that increases the power of optical signals without having to convert them

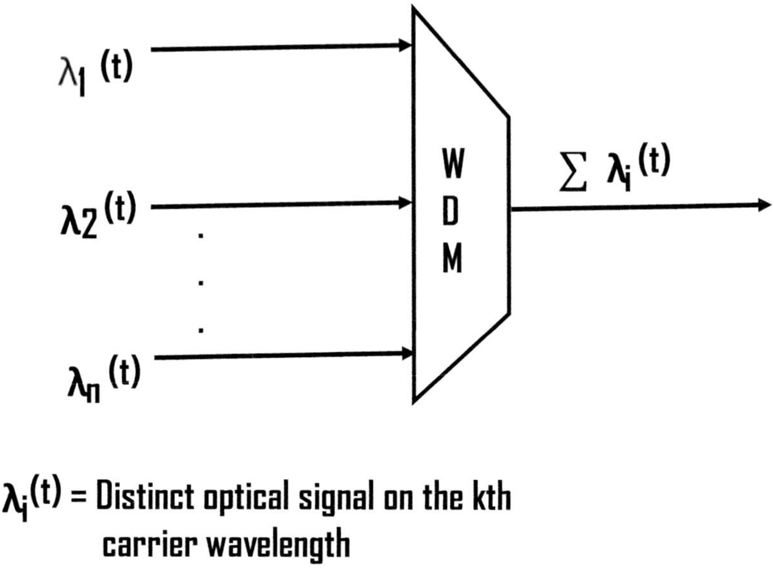

FIGURE 8.6
Optical Multiplexer.

back into electrical signals. It operates directly on the light signals, using materials such as erbium-doped fiber (EDFA). This allows longer distance data transmission and improves the overall performance of OTNs, especially in long-distance backhaul.

OFAs can be designed to operate in the 1,550 nm or in the 1,310 nm fiber transmissions windows. The operating window is determined by the fiber dopant used, i.e., erbium for the 1,550 nm window and praseodymium for the 1,310 nm window. Figure 8.7 shows an idealized gain curve for an EDFA. As indicated in the figure, the amplifier applies essentially the same gain over its complete passband.

8.1.3.1.2 Optical Add/Drop Multiplexes

An Optical Add/Drop Multiplexer (OADM) is a ONE that allows for the selective addition (insertion) or dropping (removal) of specific channels from a WDM signal without having to convert the signal back to the electrical domain. This enables efficient routing and management of bandwidth in optical fiber networks, facilitating dynamic bandwidth allocation and flexible network architecture.

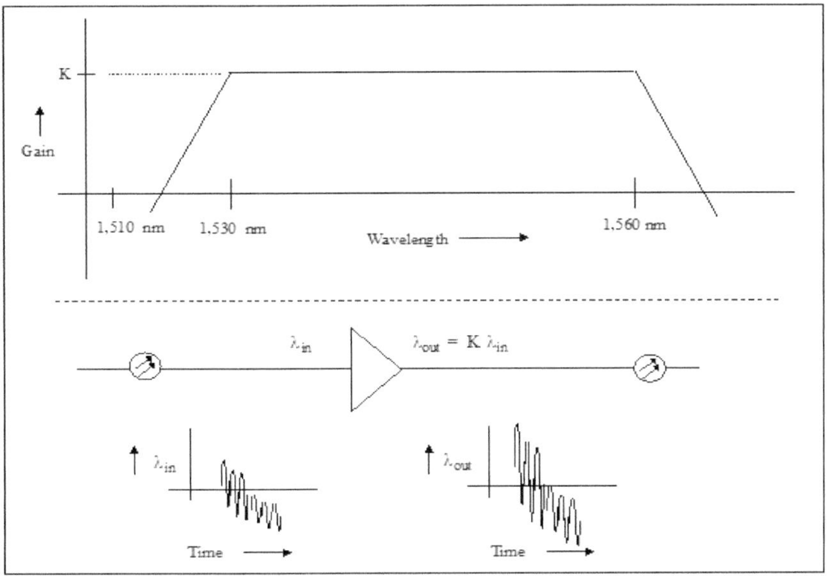

FIGURE 8.7
Optical Amplifier.

Figure 8.8 shows that the optical inputs to the WDM must operate at different carrier wavelengths.

8.1.3.1.3 Passive ONEs

Passive ONEs are optical devices that do not require an external power source. Examples of passive ONEs which may be used in telecommunications applications include:

1. Optical splitters
2. Optical connectors
3. Optical filters
4. Optical isolators
5. Optical attenuators
6. Dispersion compensators

A dispersion compensator is a component having a negative dispersion coefficient that is used to shift the zero-dispersion wavelength of conventional dispersion –unshifted fiber from 1,310 nm to 1,550 nm region. It should

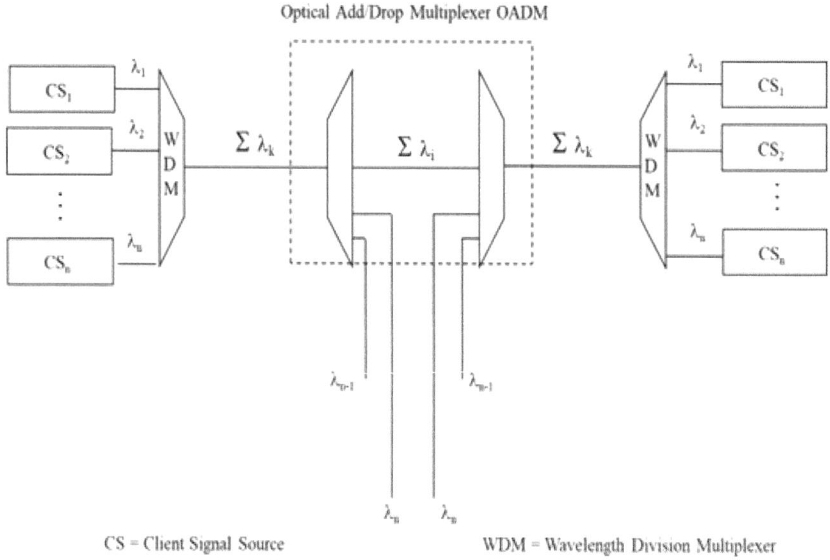

FIGURE 8.8
Optical Add Drop Multiplexer (OADM).

be noted that many of these passive Optical Network Unit (ONU) may be integrated into an active ONE.

8.1.4 OTN Topology

ITU has termed DWDM network as OTN (31, 32). OTN topology consists of three layers:

1. The optical channel layer
2. The optical multiplex section layer
3. The optical transmission section layer

Within each of these layer networks, end-to-end connection between two adjacent layers can be established. In this book, these connections are designated optical channels, optical multiplex sections, and optical transmission sections, respectively. In OTN, client server relationships exist between the optical channel layer and the optical multiplex layer and between optical multiplex layer and the optical transmission layer.

Figure 8.9 illustrates the connections between layers and end-to-end OTN transport in WDM-based network

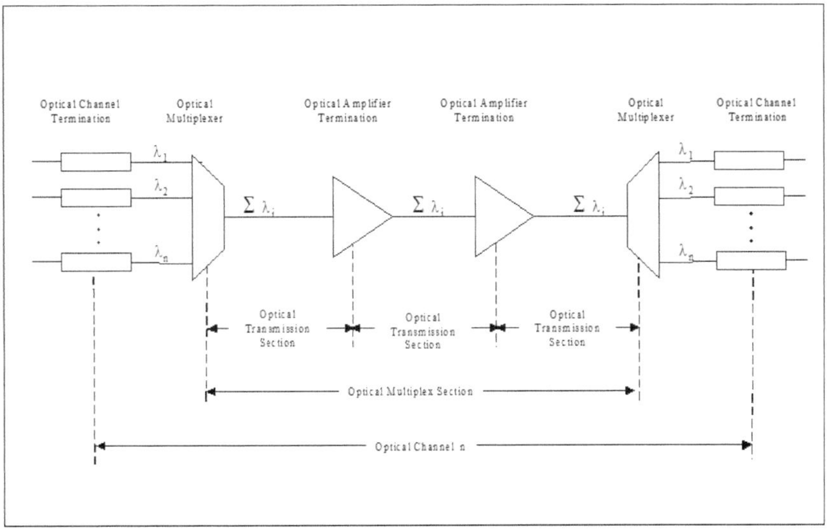

FIGURE 8.9
OTN End-to-End Transport.

8.1.5 DWDM Ring Architecture

The current OTN architecture mostly utilizes a self-healing optical ring design. A self-healing optical ring network is a type of network topology that utilizes optical fiber to create a closed-loop structure as shown in Figure 8.10. The self-healing design feature automatically performs rerouting of data in case of fiber cuts, equipment failures, or network breaks. Generally, self-healing mechanisms reroute traffic within milliseconds to seconds by quickly identifying the failure and reestablishing communication through an alternative path without manual interventions, enhancing overall reliability and resilience. The design of self-healing ring includes optical fibers and the ONEs such as Optical Amplifiers (OAs), Optical Terminal Multiplexers (OTMs), OADMs, Optical Cross Connects (OXCs), and optical transponders.

8.1.6 Multiprotocol Label Switching (MPLS)

MPLS is a protocol designed to get data packets to their destinations quickly and efficiently; it also enables the NPSBN to support IP VPN via a shared backbone. MPLS operates between the data link layer (L2) and the network layer (L3), allowing for the creation of label-based forwarding paths to improve the speed and efficiency of data transport across large and long-haul networks by enabling quick packet forwarding based on short, fixed-length

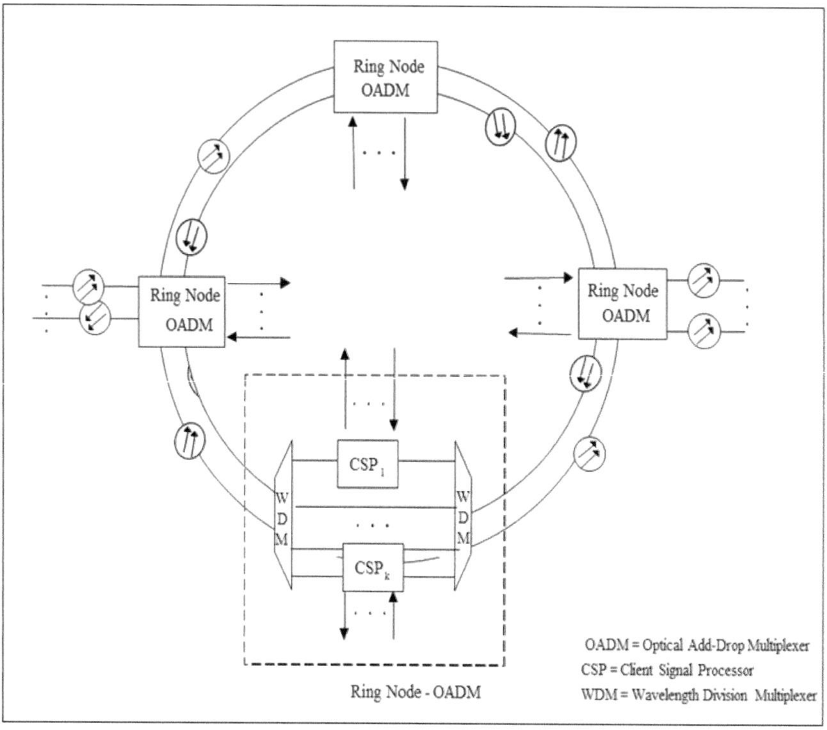

FIGURE 8.10
Self-Healing Optical Ring Network.

labels instead of network IP addresses. MPLS assigns labels to each data packet, controlling the path the data packet follows.

MPLS is an extension of the dynamic routing protocols (e.g., OSPF-TE, ISIS-TE), and fast-reroute algorithms (FRR) and other features were added to the traditional IP routers. In MPLS routing, the number of paths to route demands between nodes is not prebounded; so, it might exist as a different feasible configuration in most scenarios. OSPF uses Dijkstra's algorithm [33] to calculate the Shortest Path Tree (SPT) by creating a topological graph and using a graph search method to find the shortest path from a starting node to all other nodes in a weighted graph. It selects the node with the smallest tentative distance, updates the distances of its neighbors, and continues until it has visited all nodes. This topology is then used to populate the routing table.

MPLS is a scalable and IP-independent algorithm which greatly improves the speed of traffic, reducing network congestion by avoiding bottlenecks and optimizing traffic flow, leading to smoother and faster data transmission. MPLS enhances bandwidth utilization by allowing network administrators to prioritize diverse types of traffic. For instance, critical applications like voice and video can be given higher priority over less critical traffic. This ensures that essential services receive adequate bandwidth, reducing the likelihood of congestion and improving the network's efficiency and performance.

DWDM (OTN) combined with MPLS is a dominant network technology in high-capacity optical backbone networks. Repeaters and amplifiers must be placed at regular intervals to compensate for the loss in optical power while the signal travels along the fiber; hence the cost of OTN is proportional to its length over the physical layer. DWDM supports a set of standard high-capacity interfaces (e.g., 1, 2.5, 10, 40, 100 Gbps). The cost of a connection also depends on the capacity though not proportionally. For economies of scale reasons, the higher the bit rate, the lower the per-bandwidth cost. The client nodes together with these OTN connections form the so-called logical layer on top of the physical one.

8.2 TRANSPORT SEPARATION

FirstNet end-to-end transport is designed to support the shared BBU function to aid AT&T mobility and FirstNet traffic. AT&T's eNodeB will have logical connectivity to AT&T's commercial EPC and FirstNet EPC via shared transport network infrastructure consisting of CSBH and CBB. FirstNet traffic is carried from BBU to SIAD using VLAN-based logical connections. FirstNet traffic is routed to FirstNet-dedicated EPC based on FirstNet PLMN. FirstNet SGW/PGW transmits/receives FirstNet-specific user data from FirstNet enterprise server(s) using separate VLANs.

VLAN in communications is a way to segment a single physical network into multiple distinct logical networks. It is like creating separate lanes on the same road for diverse types of traffic to ensure they do not interfere with one another. VLANs are created by switches and segments network traffic. Each VLAN acts like a separate subnet, even though they are all part of the same physical network. Devices within the same VLAN can communicate with one another as if they were on the same physical network, but communication between different VLANs requires a Layer 3 router.

Figure 8.11 is a simplified illustration of creating four VLANs in a single physical link. Each VLAN carries a separate traffic isolated from other VLANs.

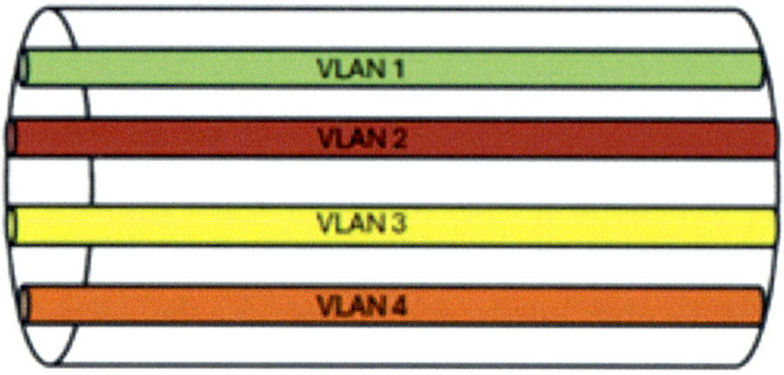

FIGURE 8.11
FN Transport VLAN Separation.

8.2.1 NPSBN Transport Protocols and Security

An IP network is a "connectionless" network transmission and each data packet in transaction is independent of the transmission of other data packets. For example, when a user accesses a website, each packet of data sent from the website is transmitted independently of all the other data packets. Connectionless networks are extremely efficient because they can optimize the routing of each packet. This necessitates a security measure in place so the data packet in transition is not vulnerable to security threats. The key security concerns for any network are listed here:

- *Authenticity:* Ensuring the source of information is accurate, that the sender of information is fully authenticated
- *Confidentiality:* Ensuring the content of data being sent through the network is not available to unauthorized personnel
- *Integrity:* Ensuring the data content is not tampered with during data transfer.

NPSBN transport network has security measures based on 3GPP/ETSI specifications (33) including encryption to protect sensitive data from being intercepted during transit. Firewalls are deployed at network boundaries to control access and filter out any malicious traffic access. Intrusion Detection/Prevention Systems (IDS/IPS) are integrated to monitor network traffic for suspicious activity and respond to potential threats effectively.

NPSBN transport security features are as follows:

- FirstNet S1 interface encrypted between eNB and Security Gate Way (SGW) in FN EPC.

- Use of Multi Operator Core Network (MOCN) to isolate FN traffic on a separate and secure VLAN
- Use of Internet Protocol Secure (IPsec) VPN in the transport segments where the S1 interface is not encrypted, e.., EPC to PDN. IPsec VPNs encapsulate the data packets, effectively hiding the original data and protecting it from unauthorized access. This encapsulation process includes both encryption of the data and the addition of an IPsec header. Encapsulation ensures data can be decrypted only by the receiving device that has the correct cryptographic key.
- CBB priority (FN traffic has end-to-end QoS across RAN, EPC, and CBB to internet egress point)
- OSPF and RIP routing protocols, static routes, IPsec secure tunnels, and QoS
- Dedicated BGP/MPLS IP VPN including IBGP and RIP routing protocols, static routes, IPsec tunnels and provisioning and activation of MPLS label switch paths
- Validation of routing configurations of IP networks and services

NPSBN OTN uses several security measures to safeguard sensitive data being transmitted using 3GPP security specifications. IPsec operates at network layer (Layer 3) that offers a comprehensive suite of protocols that provide robust security at the network layer. It creates a tunnel for transmitting data packets inside the tunnel for securing data packet transmission over data packet networks. IP tunnels are created between two IP hosts or IP routers. It ensures confidentiality, integrity, and authentication of data packets by encrypting them before transmission. IPsec also supports various encryption algorithms and key exchange methods, making it highly flexible and adaptable to different network environments. In addition, its ability to create VPNs enables secure remote access and site-to-site connectivity. During data transmission, IPsec VPNs ensure the data securityfrom unauthorized access.

IPsec VPN operations can be broken down into several key components. Initially, the VPN establishes a mutual authentication between devices at the start of a connection. This phase involves the exchange of cryptographic keys that will be used to encrypt and decrypt the data. Once authentication is complete, the devices can start transmitting data securely.

Since IPsec VPNs can be used across various hardware and software configurations, they are suitable for securely connecting disparate network segments. IPsec VPNs create a secure, reliable method of communication which creates a user experience like being directly connected to the enterprise network.

8.2.2 VPN (Virtual Private Network)

A VPN encrypts data and creates secure connections between remote users or locations between the private PDNs or between public and private PDNs, thereby ensuring privacy and security while transmitting sensitive information across potentially insecure environments. VPN consists of the following two components:

- Edge network equipment that provides access to the IP backbone. This may be customer premises equipment (CPE) or a network edge router.
- Secure tunnels through the IP backbone. A tunnel is a means to connect two points through an IP backbone.

In addition, there are security measures covering only specific internet sessions connecting to an application. Hypertext Transfer Protocol Secure (HTTPS) is a protocol that protects the data being exchanged between a user and applications, e.g., paying a bill over the internet. HTPPS sessions provide security only for the life of the active session and do not extend once the session is over. HTTPS uses Transport Level Security (TLS) to secure communication between a web browser and a web server at the application layer.

8.2.3 NPSBN OTN Performance QoS

NPSBN's end-to-end OTN comprises the following intermediate node transport:

1. eNodeB – SIAD
2. SIAD – Metro Transport (CSBH)
3. CSBH – MTSO/MSN
4. MTSO/MSN – Backhaul (CBB)
5. CBB – S/PGW
6. S/PGW – PDN

Figure 8.12 provides the expected transport time for each segment. It should be noted that the transport time between S/PGW and PDN is not mentioned

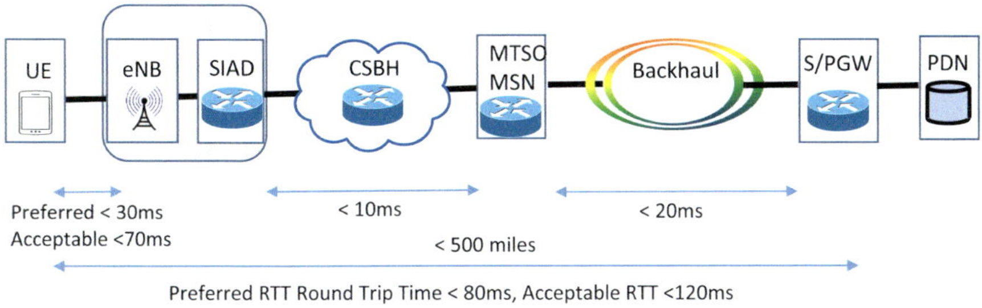

FIGURE 8.12
NPSBN OTN End-to-End Transport Performance.

since it is generally not part of the NPSBN but owned and managed by enterprises

8.2.3.1 NPSBN CSBH Transport QoS

NPSBN CSBH transport QoS parameters are described in Table 8.1 which is based on the MEF 10.2 specifications document of the Technical Committee of the MEF that specifies the Ethernet service attributes (34).

8.2.3.2 NPSBN End-to End Class of Service (CoS)

An end-to-end priority and marking scheme Differentiated Service Code Points (DSCP) is designed to provide a consistent Class of Service (CoS) treatment across RAN, CSBH, CBB, EPC, and USP/IMS networks for FirstNet subscribers. RFC 2474 (35) specifies the Differentiated Services Field in IPv4 and IPv6 headers, known as the Traffic Class field. It establishes a framework for providing different levels of service in IP networks, allowing for better traffic management and performance.

RFC 2597 (36) defines additional features of Assured Forwarding (AF) and Expedite Forwarding (EF) for DSCP. AF guarantees a certain amount of bandwidth, Guaranteed Bit Rate (GBR), and even allow access to extra bandwidth, if available. EF is used for low loss, low latency, low jitter, assured bandwidth, and end-to-end service appearing as point-to-point connection mainly used for "premium services." This is crucial in NPSBN for ensuring QoS across several types of services, particularly in handling voice, video, and data traffic efficiently.

TABLE 8.1

CSBH Transport Quality of Service

Performance Attributes	Transport Class		
	Real Time	**Priority Data**	**Normal Data**
Mobility Applications	QCI 1-3	QCI 4, 5, 7	Q.CI 6, 8, 9
Mean Frame Delay (ms)	≤5	≤13	≤20-28
Frame Delay Range (ms)	≤3	≤10	≤20
Frame Loss Ratio	<5 x 10E-5	<10E-4	<10E-3
Convergence	<3s	<3s	NA
Availability	Cell Site: 99.99%, MSN link pair: 99.999%		

Notes:

1. All parameters shown are for "one way."
2. CSBH performance attributes for *latency* (i.e., Mean Frame Delay), *jitter* (i.e., Frame Delay Range), and *loss* (i.e., Frame Loss Ratio) described follows MEF specifications. Current Single Class (RT) Ethernet transport performance requirements are (a) Frame Delay: < 30 ms, (b) Jitter: < 3 ms, (c) Frame Loss Ratio: < 10E-6, and (d) availability: 99.99% for Cell Site, and 99.999% for MSN pair-links.

TABLE 8.2

NPSBN DSCP and PMLS EXP Assignments for FN Services

Application Type	FirstNet Traffic on NPSBN	
	QCI – DSCP	**MPLS Priority**
VoLTE	QCI 1 (DSCP 46/EF)	Real Time, EXP 5
SIP	QCI 5 (DSCP 32/CS5)	Private, EXP 4
eMBMS	QCI 4 (DSCP 34/AF41)	Private, EXP 4
Mobile Data – High	QCI 6 (DSCP 18/AF21)	Default, EXP 2
IR.94 Video	QCI 8 (DSCP 22/AF23)	Default, EXP 1
Mobile Data – Normal	QCI 8 (DSCP 22/AF23)	Default, EXP 1
Mission Critical	QCI 65 (DSCP 40/CS5)	Real Time, EXP 5
	QCI 66 (DSCP 46/EF)	Real Time, EXP 5
	QCI 69 (DSCP 41)	Real Time, EXP 5
	QCI 70 (DSCP 26/AF31)	Private, EXP 4

MPLS EXP (MPSL Expansion) in the OTN is a framework that focuses on the efficient management and scalability of layered communication services over optical networks. It emphasizes resource allocation, performance optimization, and the integration of different service layers, enhancing the overall capacity and flexibility of optical transport systems. Priority markings across the various network domains and the transport network across the

FN network to provide an end-to-end QoS for various FirstNet services are provided in Table 8.2.

8.3 NPSBN TRANSPORT NETWORK FOR 5G

5G transport requires Ultra Low Latencies (uRLLC) transport to support vertical domains (37) including enhanced Mobile Broadband (eMBB), Machine to Machine (M2M) communications, self-driving vehicular system communications, Vehicle to Everything (V2X) communications, intersatellite communications, etc. 5G transport technology in NPSBN will use the shared OTN currently deployed for NPSBN LTE/4G transport. The requisite enhancements for 5G network transport are still in the evolving process with the standards communities. It is expected that 5G technology using Millimeter Wave (mmWave) spectrum, efficient edge computing, efficient computing, etc. will require an enhanced OTN to take full advantage of the 5G. Edge computing is a technology that processes data closer to where it is generated, rather than relying solely on centralized data centers. This approach reduces latency, enhances speed, and improves efficiency, making it ideal for applications like Internet of Things (IoT) devices, smart cities, and real-time data processing.

8.3.1 NPSBN Transport Security

NPSBN transport security measures protect data during transmission. This includes protocols and mechanisms that ensure the confidentiality, integrity, and authentication of the transmitted information over the air interface and underneath the LTE network infrastructure.

The key components of LTE transport security include:

1. Encryption: To protect user data and signaling messages from unauthorized access.
2. Integrity Protection: To ensure that the data has not been altered during transmission.
3. Authentication: To confirm the identity of the device and the base station.

8.4 NPSBN TRANSPORT – KEY POINTS

NPSBN end-to-end transport spans from user devices to user application data, e.g., PDN, internet, public safety databases. It is divided into following segments based on the specific purpose and technology:

- *Radio Wave:* This is between UE and the 4G/5G RAN.
- *Front-haul:* These are transport segments between 4G/5G RAN and SIAD. A SIAD is an aggregation node connecting to several eNBs/gNBs. Generally, it is a fiber-based transport segment.
- *CSBH:* This is a Metro Ethernet–based transport segment connecting SAID to MSN in an MTSO. The physical layer includes cable or fiber.
- *Backhaul:* It is the major segment of the NPSBN transport and spans between MTSO and the EPC offices. It typically involves high-capacity transmission links that carry the data from UE/RAN/SIAD to the core network. NPSBN backhaul, called CBB, uses MPLS technology. MPLS is based on the self-healing optical ring design and routing data from one node to the next based on short path labels rather than long network addresses.
- *Mid-haul:* It is a transport segment spanning between backhaul and the LTE EPC centers. Generally, it is point-to-point or multipoint connections over fiber cable connecting and backhaul/MPLS to specific EPC node.

NPSBN transport security measures provide the following vital data security during data packet transport:

- *Authenticity:* Ensuring the source of information is accurate; that the sender of information is fully authenticated.
- *Confidentiality:* Ensuring the content of data being sent through the network is not available to unauthorized personnel.
- *Integrity:* Ensuring that data content is not tampered with during data transfer.

NPSBN security measures include IPsec/VPN between UE and eNB/gNB. In addition, S1 (S1 U and S1 C) interfaces are encrypted between eNB/gNB and SGW in FN EPC which provide data encryption in front-haul, CSBH, backhaul, and mid-haul transport segments. Data packet transmission from

EPC to PDN is safeguarded using VPN which provides authentication, confidentiality, and data integrity.

NPSBN transport prioritization includes QoS parameters in different segments of the end-to-end transport. CSBH performance attributes include frame delay (latency), frame delay range (jitter), and frame (packet) loss. MEF specifications: (a) latency <30 ms, (b) Jitter <3 ms, and (c) Packet loss <10E-6. Network availability: 99.99% for Cell Site and 99.999% for MSN.

NPSBN transport meets RFC 2474 Differentiated Service Code Points (DSCP) as well as MPLS EXP for FirstNet services.

9

The FirstNet Services

The U.S. Congress mandated the development of National Public Safety Broadband Network (NPSBN) for the first responders to meet their public safety commitments to save lives and properties in all U.S. states and territories around the clock effectively and efficiently. This chapter describes the multitude of services developed for FirstNet (FN) users/first responders which are supported by NPSBN.

It should be taken into consideration that although the first responders are the focus of the advent of NPSBN services, the public safety at large involves many other users who get involved in public safety incidents and assist first responders. The FN users are drawn from different functional areas, could be part- or full-time employees, employees from utility's companies, hospitals, or volunteers.

FN NPSBN has dedicated spectrum and Evolved Packet Core (EPC) but share other network resources with FN private partner American Telephone and Telegraph Company (AT&T) to provide network services to FN users. Figure 9.1 illustrates the different users sharing the NPSBN/AT&T network resources.

9.1 FN USERS

FN users are classified based on their role-plays in public safety operations and management as well the organizations they belong to. The type of service, network access priorities, Quality of Service (QoS), etc., vary for different

DOI: 10.1201/9781003557739-11

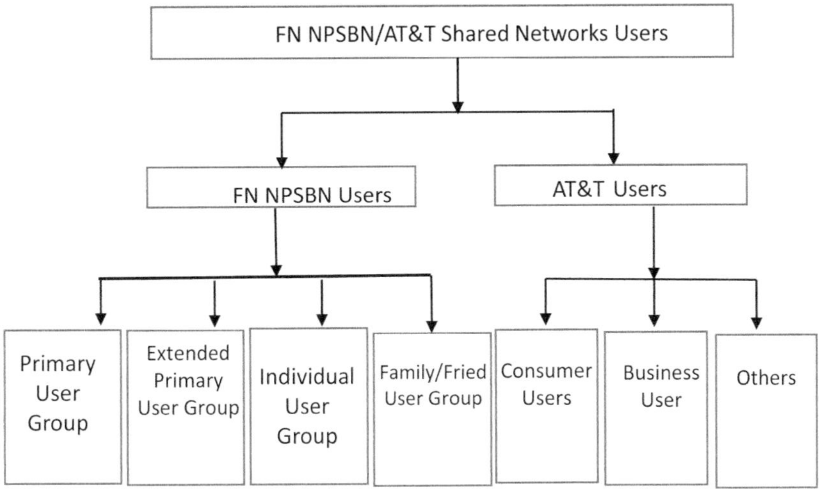

FIGURE 9.1
NPSBN and AT&T – Shared Network Users.

user groups in accessing and using NPSBN services. The following are the user groups that can access NPSBN and avail FN services.

- Primary user group (PU)
- Extended primary user group (EPU)
- Individual responder group
- Family and friends – With a FirstNet unlimited smartphone plan line of service (minimum $39.99/month) and an AT&T unlimited plan line of service (minimum $75/month without discounts)

9.1.1 FN Primar User Group

Primary users are the employees of the Public Safety Entities (PSEs) such as police, fire, emergency medical service (EMS), 9-1-1, emergency management, etc. Police, fire, and EMS users are also called first responders because they provide immediate and first assistance during emergencies or disasters. The primary users have access to NPSBN services with their FN-enabled devices and have Quality of Service (QoS), priority, and preemption (QPP) capabilities on their service profile enabling them to attach to NPSBN and perform their duties.

9.1.2 Extended Primary User Group

Although, initially, the intent of the FN NPSBN was use only by primary users/first responders, soon it was realized that many other organizations and services – as well as critical infrastructure – are often needed during emergencies and in the aftermath of a catastrophic event. Extended primary users are drawn from both public and private organizations – mainly from electric utilities and power grid companies, telecommunication services, hospitals and medical facilities, schools, water supply and waste management companies, transportation systems including airports, public transit, and rail system, construction companies, sanitation departments, etc., whose services could be required during emergencies such as building fires, hurricane, floods, wild fires, and building collapse. Extended users are involved to perform certain public safety functions and to assist in specific emergencies.

9.1.3 Individual Responder Group

Individual users are generally not employees of any public safety agencies but are self-employed consultants, volunteers, etc., who may be asked to join the rescue operations or similar public safety challenges in an incident by the duly designated incident manager. These users can access the NPSBN services only for the duration of their involvement during the incident management.

The incident manager who is official-in-command will authorize the individual responder access to the NPSBN along with the level of QoS and other features. Table 9.1 provides the service eligibility of different user groups accessing FN NPSBN.

FirstNet users have priority over non-FirstNet users and, if necessary, all commercial traffic will be dropped and the network will be dedicated

TABLE 9.1

FN NPSBN Subscriber Profile on SOC

FirstNet QoS, Priority, and Preemption SOC by User/Subscriber Profile for FirstNet SIM		
Primary User	**Extended Primary User**	**Individual User (Subscriber Paid)**
FNPRIM2 (Platinum)	Not Available	Not Available
FNPRIM3 (Gold)	FNEXTPR4 (Silver)	Not Available
FNPRIM4 (Default-Silver +)	FNEXTPR (Default)	SPFNPRIM4 (Default)

exclusively to first responders, as well as to the EPU group as bandwidth allows. During a crisis, a normal cellphone might stop working, but FirstNet-enabled devices will continue to work.

9.2 SUBSCRIBER PROFILE IDENTIFIER

Subscriber Profile Identifier (SPID) is a code assigned to subscribers in Home Subscribers Service (HSS). SPID defines idle mode priorities and controls inter-Radio Access Technologies (RAT) or interfrequency band handover in connected mode. At the time of user equipment (UE) setup/modification, Mobility Management Entity (MME) sends subscriber's SPID over S1 interface to eNB which maps SPID to a local configuration list to select radio resources.

SPID and interfrequency load balancing (IFLB) features make B14 the preferred band for FirstNet subscribers. SPID mechanism steers non-FN traffic away from B14, especially during network congestion. All non-FirstNet subscribers will be assigned a SPID value in the HSS. The eNB uses this value to determine the frequency band priority for these users. B14 will either not be listed or be the lowest priority for all non-FN subscribers.

Table 9.2 shows a sample SPID assigned to FirstNet PU and EPU and along with provisioning requirement on HSS.

9.3 NPSBN SERVICE REQUEST GENERAL FLOW

A generic NPSBN service request flow is presented in Figure 9.2.

TABLE 9.2

SPID Assignment Example

SPID	User Group	HSS	Bands Prioritization Example
1	FirstNet primary users	Y	14, 12, 2, 4, 5, 30 ...
2	FirstNet extended primary users	Y	14, 12, 2, 4, 5, 30...
Blank	AT&T commercial users	N	12, 2, 4, 5, 30, 14

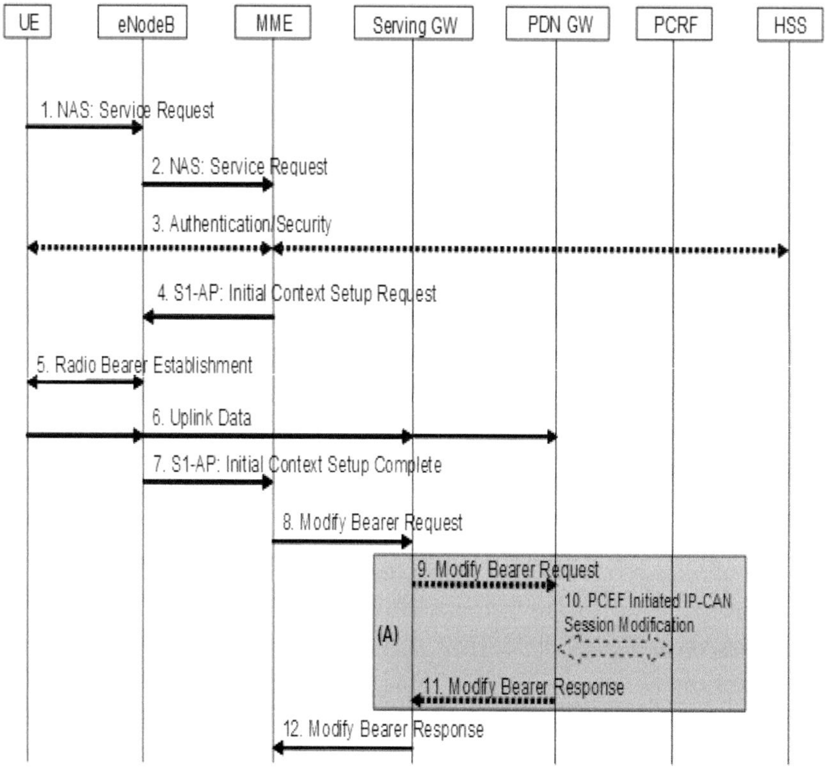

FIGURE 9.2
NPSBN Generic Service Request Flow.

9.4 VOICE OVER LTE SERVICE

NPSBN Voice over LTE (VoLTE) is a High Definition (HD) voice calling service over the 4G/5G wireless networks. The readers are familiar with the phone calling services provided by Public Switched Telephone Network (PSTN) either by circuit switches or voice over an IP (VoIP) network.

HD voice calling over Long-Term Evolution (LTE) features include:

- HD voice including call forwarding, call waiting, and multileg
- Unstructured Supplementary Service Data (USSD)-based service (e.g., Star Services – *55 for road condition)
- N11 dialing – three-digit number used for specific purpose, e.g., 311 used in New York City for non-emergency city services

- Abbreviated dialing – calling with short digit sequence to reach specific telephone numbers rather than dialing full phone number
- Directory assistance
- 611 customer service
- Hot lining
- Visual Voicemail (VVM) with Message Waiting Indicator (MWI)
- Regulatory required services such as Teletypewriter (TTY)/Hearing Aid Compatibility (HAC)
- Simultaneous voice and data

9.5 VOLTE SERVICE ARCHITECTURE

Figure 9.3 shows a high-level logical VoLTE service architecture with Radio Access Network (RAN), FN EPC, IP Multimedia Subsystem (IMS) core, and Converged Telephony Server/Application Server (CTS/AS). VoLTE uses Session Initiation Protocol (SIP). The calling party IMS core connects to the called party IMS core and establishes the connection after the call is answered. Subsequently, voice data packet starts in the form of voice conversation. Either party drops the call after the call is over.

NPSBN VoLTE interfaces with PSTN for landline number calls. Connecting to PSTN is performed using Circuit Switched Fallback (CSFB) feature.

9.5.1 VoLTE Service Design and Call Flow

VoLTE call flow involves the steps to initiate and complete a call in an LTE network. The typical VoLTE call flow involves several key processes:

1. Registration: The user's device registers with the LTE network, confirming that it supports VoLTE.
2. Call Setup: When a user makes a call, the device sends a request to the LTE network using SIP.
3. Session Establishment: The IMS initiates the call setup. The network locates the recipient's device and establishes a media session.
4. Real-Time Protocol (RTP) Stream: Once the call is accepted, an RTP stream is created for transmitting voice packets.
5. Call Progression: The call proceeds, allowing real-time voice communication between the calling party and the called party.

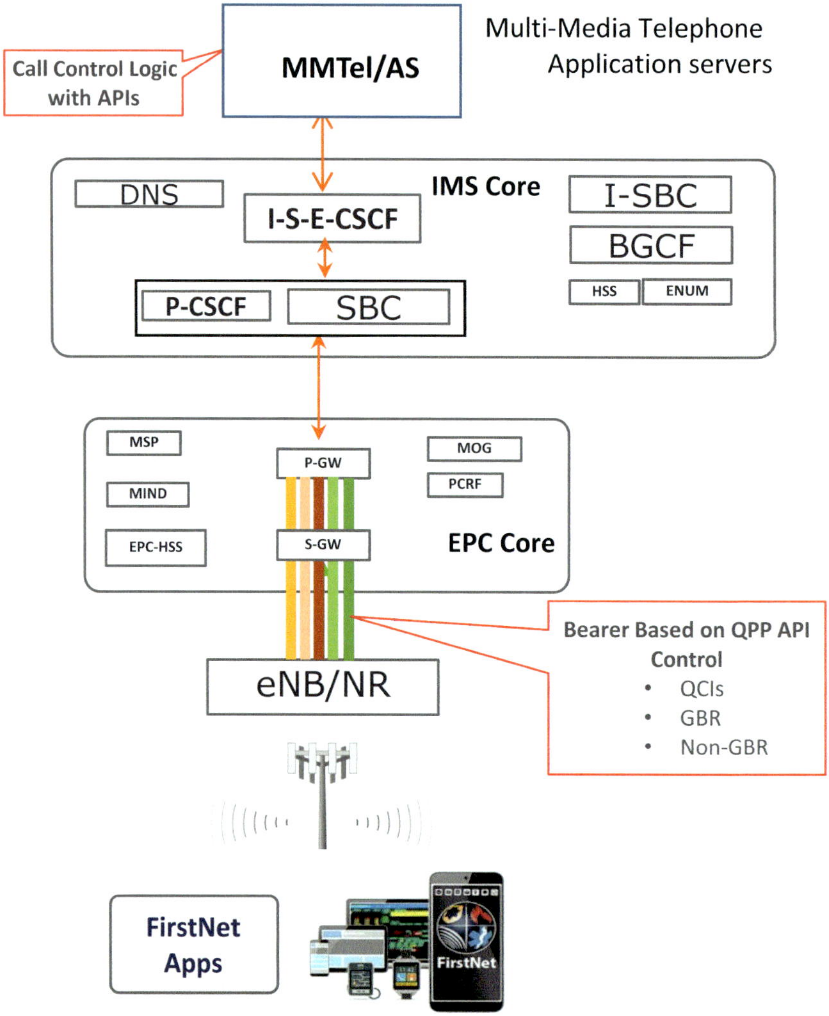

FIGURE 9.3
NPSBN VoLTE Service Logical Architecture.

6. Call Termination: After the conversation, the call is terminated, and the network is informed to release the resources.

A simplified VoLTE call flow is given in Figure 9.4. The calling party initiates a VoLTE call and the calling party UE sends SIP message to the called party UE. When the calling party rings and is picked up, the VoLTE bearer is established, and voice stream is exchanged between UE via Packet Gateway (PGW).

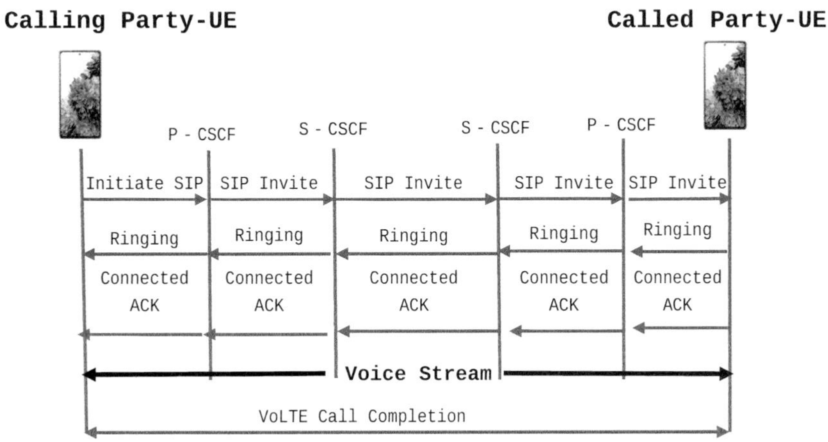

FIGURE 9.4
VoLTE Call Flow.

9.5.2 VoLTE Voice Performance and QPP

HD VoLTE quality is highly dependent on the network latency parameter. It is desirable that NPSBN VoLTE voice quality should be at par or better than PSTN voice quality. Figure 9.5 provides the factors that impact voice quality.

Mean Opinion Score (MOS) is a widely used measure of voice quality in telecommunications and ranges from 1 to 5, with 5 being the best quality and 1 being the worst. MOS is derived from subjective evaluations where listeners rate the quality of a voice signal, often influenced by factors such as clarity, distortion, and background noise. The score helps assess user experience for voice communications.

Adaptive Multi-Rate Wideband (AMR-WB) is a codec used for compressing voice audio, designed to deliver high-quality voice communications with improved intelligibility, especially in noisy environments. It supports a wider audio frequency range compared to traditional narrowband codecs, resulting in clearer and more natural-sounding speech. The bandwidth of AMR-WB extends up to 12.65 kHz, allowing for better handling of voice nuances and a more pleasant listening experience.

Rating Factor (R-Factor) is one of the two most used measurements of overall voice call quality in IP networks. The R-Factor value is derived from metrics such as packet loss, packet delay variation (jitter), and latency. It allows you to quickly assess the quality of experience for calls on your VoIP network. R-Factor is a key metric used in voice quality measurement systems to quantify the perceived quality of voice communications, particularly in

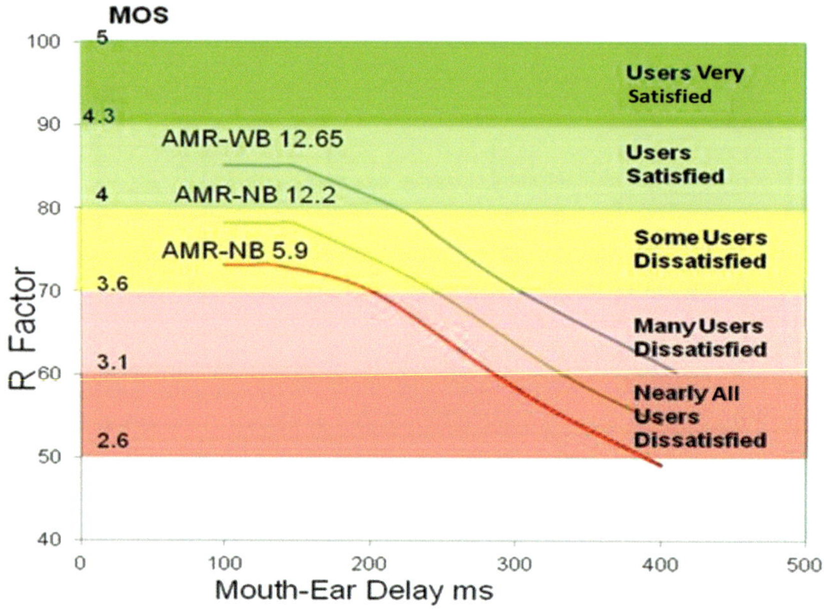

FIGURE 9.5
Factors Affecting Voice Quality.

VoLTE networks. It combines several factors, including the MOS, for a comprehensive evaluation. The R-Factor ranges from 0 to 100, with higher values indicating better voice quality. It considers aspects such as codec type, packet loss, delay, and jitter to provide an overall assessment of voice transmission quality. These factors prove that NPSBN VoLTE quality may vary while the user is mobile.

9.5.3 VoLTE Service – Key Points

Voice over LTE (VoLTE) is the most used wireless service. NPSBN provides a HD voice calling service over the 4G/5G wireless networks and supports most of the services provided by VoIP or PSTN.

9.6 IP MESSAGING (IPME)

IP Messaging (IPME) in NPSBN uses internet protocol (IP) technology to send and receive messages over the LTE network rather than relying on traditional Short Message Service (SMS). IPME uses Rich Communication Services (RCS) and unlike SMS, IPME operates across different platforms and devices, making it a more versatile and an enhancing user experience. NPSBN conforms to RFC 6665 specifications published by the Internet Engineering Task Force (38).

NPSBN IPME enables users to send text, images, videos, and more through dedicated applications that utilize mobile data or Wi-Fi, providing features like group messaging, reading receipts, and typing indicators.

9.6.1 IPME Service Design

NPSBN IPME service design differs in Apple devices and devices using Android IOS. Apple has integrated messaging systems and support MMS whereas Android-based devices use RCS client as a downloadable client. FN users can send Multimedia Services (MMS) and transfer files to all FN devices. FN users can send MMS to one user or several users, though the maximum number of recipients are limited to less than 10 (RFC 6665).

9.6.2 IPME Service Performance and QPP

NPSBN IPME Service complies with 3GPP and IETF specifications.

9.6.3 IPME Service – Key Points

Messaging service is an important NPSBN service feature which is available in both Apple iOS and Android iOS-based devices. IPME supports both text and picture transmission using Rich Text Communication (RTC) whereas the older SMS was limited to short text.

9.7 IR94 VIDEO CALLING SERVICE

IR94 video calling service facilitates video calling over a 4G network using the IMS. It allows users to make high-definition video calls, taking advantage of the enhanced bandwidth and low latency provided by 4G technology. NPSBN IR94 is based on the Third Generation Partnership Project (3GPP) standards specifications for IR94 calling (39). These specifications outline the requirements for IMS-based RCS, which includes voice and video

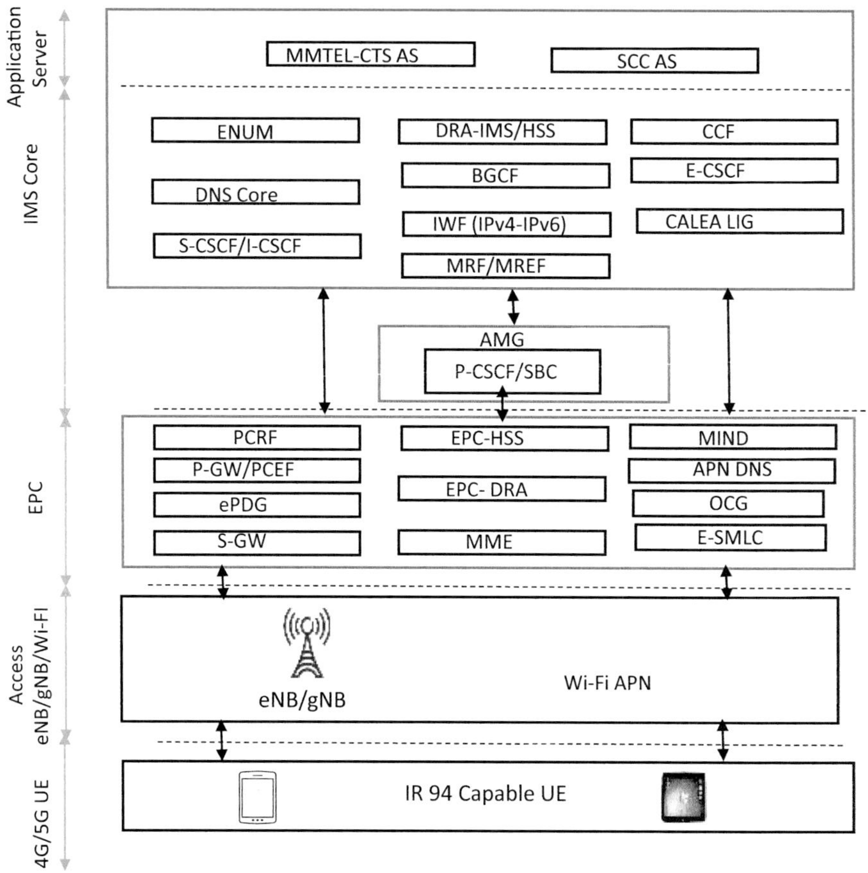

FIGURE 9.6
NPSBN IR94 High-Level Logical Architecture.

calling features. The IR94 standard specifically focuses on the interworking of conventional voice services with IMS, enabling improved call experiences through enhanced functionalities and services.

Figure 9.6 outlines the logical architecture of the IR94 video calling service, focusing on its integration within the IMS framework to ensure high-quality, interoperable conversational video and voice communication.

9.7.1 IR94 Video Calling Service Design

IR94 video calling service utilizes VoLTE architecture. Functionally, VoLTE architecture can be divided into three categories, i.e., UE, Evolved Packet System (EPS)-LTE Access, and Universal Service Platfor/IP Multimedia

Subsystem (USP /IMS) core. Similar to VoLTE call, IR94 includes IMS features, e.g., SIP registration procedure, RTP profile configuration, and call establishment and termination for video calls. Also, bearer assignments for video packet capabilities are established. NPSBN QPP are applied based upon the FN user profiles.

9.7.2 IR94 versus Over the Top Video Calls

1. Standardization and Compliance:
   - IR94: NPSBN IR94 conforms with IMS framework defined by 3GPP specifications which ensures interoperability and high-quality service across different wireless devices and networks. 3GPP specifications are natively integrated into the IMS to support IR94.
   - OTT Applications: These applications, such as FaceTime, Skype, and others, operate independently of network operators and are not bound by the same regulatory standards. They may vary widely in terms of features, quality, and interoperability.
2. Integration with Network Services:
   - IR94: IR94 is integrated into the NPSBN network, leveraging VoLTE capabilities, which allows for enhanced real-time communication and interaction, including HD voice and video calls.
   - OTT Applications: These services rely on the internet and do not utilize the underlying network infrastructure in the same way, which can lead to differences in performance and reliability.
3. QoS:
   - IR94: This is designed to support FN QPP and guaranteed QoS, full-duplex voice, and video with tight synchronization, ensuring a seamless user experience during point-to-point and multiparty calls.
   - OTT Applications: While many OTT services offer good quality, they can be affected by network conditions, leading to variable performance. They do not have the same level of QoS guarantees as IR94.

In summary, IR94 highlights the importance of compliant features as defined by 3GPP standards and compares the IR94 specifications with over-the-top (OTT) video chat applications. IR94 is a standardized, network-integrated solution designed for high-quality video calling, while OTT applications offer more flexibility and consumer-oriented features but may lack the same level of service quality and interoperability

9.7.3 NPSBN IR94 Service – Key Points

- The IMS profile for conversational video service sets mandatory features for interoperability between wireless devices and networks.
- Conversational video services include full-duplex voice and variable video streams accommodating both point-to-point and multiparty conference calls.
- The Global System for Mobile Association (GSMA) IR92 and IR94 specifications are preferred standards for deploying voice and video services over LTE networks, meeting industry expectations.
- The IR94 standard plays a crucial role in enabling remote interactions between individuals and machines by leveraging advanced network capabilities.
- NPSBN IR94 provides FN QPP, resulting in high-quality voice and video.
- IR94 described for LTE/4G is equally compatible with 5G and requires no additional features.

9.8 WI-FI CALLING SERVICE

FirstNet Wi-Fi calling (voice and video) service, also referred as Wi-Fi off-load, provides additional connectivity options for first responders in the event the 4G/5G radio network is down, not available, or below acceptable threshold and public Wi-Fi is accessible. Wi-Fi off-loading usually means the "Wi-Fi off-load (handover)" between LTE and Wi-Fi networks.

The Wi-Fi calling end-to-end system initial authentication is based on Extensible Authentication Protocol-Authentication Key Agreement (EAP-AKA) authentication via the SIM's cryptographic key (Ki) value. FN SIM contains a Ki, which is also stored in the HSS. This key is used to have the UE authenticate the network. An EAP token is generated and used for seven days to minimize Authentication, Authorization, Accounting (AAA)/HSS queries. Service tokens and IMS credentials are given only to the "primary device" based on authentication via EAP-AKA or EAP token. Once the authentication procedures are complete and a session is established, a new IMS APN IPsec tunnel is established between the UE and Enhanced Packet Data Gateway ePDG, as well as a new GTP tunnel between v-PDG. NPSBN PGW and vEPDG enable SIP registration from the device to the IMS core.

While SMS and MMS delivery are allowed over Wi-Fi, those services are protected within an IPsec tunnel that is authenticated by the SIM's Ki values.

9.8.1 Wi-Fi Calling Service Architecture

Figure 9.7 provides a Wi-Fi service flow diagram or a high-level architecture.

9.8.2 Wi-Fi Calling Service Design

1. UE activates 802.11 protocols with WLAN protocol to connect to Wi-Fi APN.
2. IP is allocated for UE Wi-Fi using proxy service
3. ePDG discovery/attachment process – UE queries APN Domain Name System (DNS) for FN ePDG and receives ePDG address in IPv4 or IPv6 of NTC nearest to UE. UE sets up an IPsec tunnel.

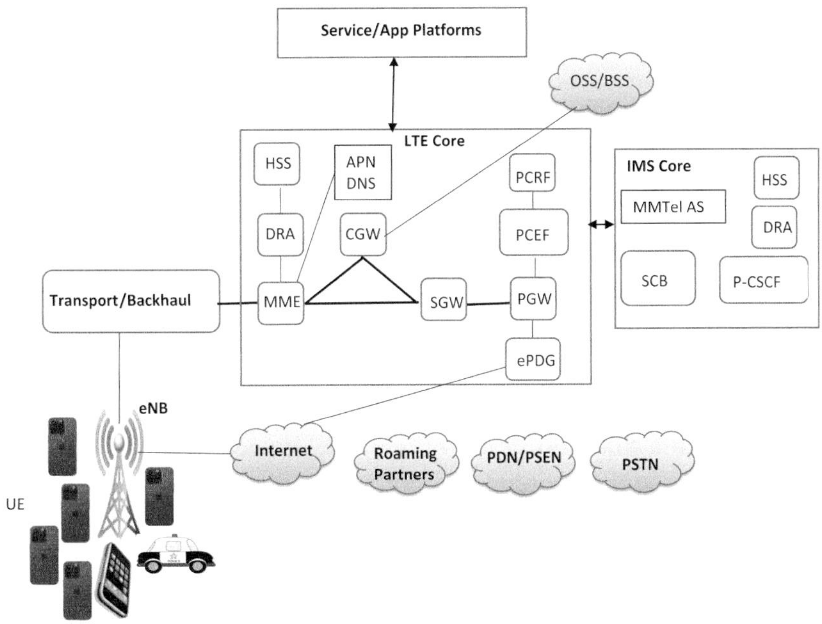

FIGURE 9.7
NPSBN Wi-Fi Calling Service Design Architecture.

4. The ePDG queries APN-DNS to find the PGW IP address. vEPDG provides a list of APN DNS servers and queries based on its geographic location. vEPDG queries the APN DNS server in a round-robin fashion for the PGWs with the desired application APN – IMS APN, NXTGENPHONE APN, SOS APN, etc.
5. APN DNS provides a list of the PGW. EPDG performs topological matches with its location and selects the closest geographical PGW.
6. vEPDG creates the session with the selected PGW.
7. Communication starts using IPsec tunnel.

The following services are available over Wi-Fi network using ePDG.

- IMS voice
- Messaging – SMS over IP, MMS
- Supplementary services (i.e., call forwarding, call waiting)
- VVM (Visual Voice Mail)
- Video calling (IR94)
- Secure access provided via dedicated vEPDG.

Seamless handoff: It is important to note the interoperability/handover between VoLTE and Wi-Fi is possible only when Wi-Fi network is already connected.

- Seamless handover from VoLTE to Wi-Fi and Wi-Fi to VoLTE
- No handover between Wi-Fi and circuit switch (PSTN)
- No handover between Wi-Fi to another Wi-Fi. Handover is supported if Wi-Fi has same Service Set Identifier (SSID), which is a network, and same IP is maintained. If the IP address changes, the call will drop.

9.8.3 Wi-Fi Calling Service QPP and Performance

FN QPP is not available over Wi-Fi network. However, the data packet is secured within the IPsec tunnel.

9.8.4 Wi-Fi Calling Service – Key Points

NPSBN Wi-Fi handover is an important feature for first responders for continuity in critical communications. IPSec provides security for data packet transmission between Packet Data Network (PDN)/applications and UE.

9.9 NPSBN DATA SERVICES

In LTE network, data service provides efficient transmission of data and connectivity to first responders for internet and application access. It utilizes high-speed packet data and transmission methods enabling users to access various data services such as web browsing, video streaming, uploading (UL), and downloading (DL) including access to private data networks. NPSBN network provides rigorous data security and performance specified by 3GPP while transmitting data and improves overall network efficiency by employing technologies such as Multiple Input Multiple Output (MIMO), QPP, data size fragmentation, etc.

9.9.1 NPSBN Data Service Architecture

NPSBN data service architecture is given in Figure 9.8, which is a short view of the LTE core architecture. Data packet transmission starts after the UE initial authentication and attachment are completed. It involves Serving Gateway (SGW) and Packet Gateway (PGW). FN QPP is enforced by Policy and Charging Enforcement Function (PCEF) after QPP profile is verified in Policy and Charging Resource Function (PCRF).

S1 control traffic is managed by MME which also interfaces with SGW. SGW selects the PGW based on the APN to attach PDN over internet and send S1 user data to UE via eNB. eNB also interacts with neighboring eNBs via X2 interface and communicates with MME or SGW appropriately during the UE movement depending on the signal strength.

9.9.1.1 Data Packet

The NPSBN network uses a packet-switched architecture allowing for efficient transmission of data by breaking it into smaller manageable pieces at the start of the transaction which are then sent over the network and reassembled at the receiving end.

A data packet consists of several key components:

1. Header: Contains metadata used for routing and delivery, such as source and destination IP addresses, protocol type, and sometimes time-to-live (TTL).
2. Payload: The actual data being transmitted, which can include application-level data.

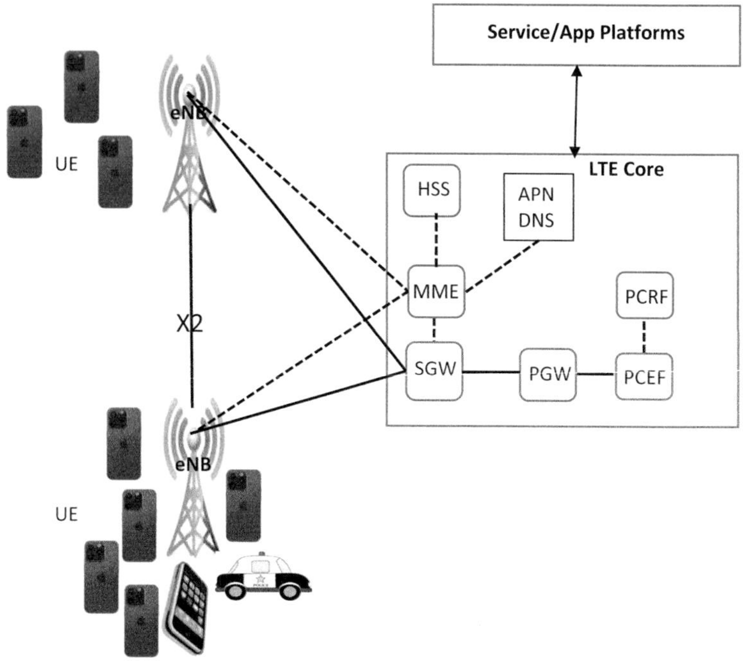

FIGURE 9.8
NPSBN Data Service Logical Architecture.

3. Footer (or Trailer): Often includes error-checking information, such as checksums or frame delimiters, to ensure data integrity on arrival.

9.9.1.2 Size of Data Packet

NPSBN allows the maximum data packet size called Maximum Transitable Unit (MTU) of 2K bytes. 3GPP TS 36.213 (40) provides LTE data packet size recommendation. 3GPP 36.213 is a technical specification from the 3GPP that covers the physical layer procedures for LTE in the RAN. It mainly focuses on aspects like the modulation schemes, error correction coding, and transport format combinations relevant to the physical layer. NPSBN SAE-GW IPv6 RA MTU is set to maximum 2,000 bytes by default. It means that packet size being transmitted will have the maximum packet size of > 2,000 bytes which include header, payload, and footer bytes. The MTU defines the largest size of a single data packet that can be transmitted over the network, which encompasses all associated headers and the actual data (payload) being sent.

NPSBN has additional data security requirements over the standards LTE transmission security essentially required for all public safety data transmission by incorporating S1U encryption. For S1 U encapsulated data packet MTU will be changed from 2,000 bytes to a lower value > 1,320 bytes.

Radio Link Control (RLC) protocol operates at the Data Link Layer (Layer 2) of the LTE protocol stacks and is responsible for managing the reliability and flow control of data transmissions between the UE and the eNodeB (base station). RLC protocols perform data segmentation by breaking higher layer data packets into RLC data packet unit (PDU) for transmission over radio interface. RLC reassembles the PDUs into complete data packet at the receiving end. Also, RLC performs concatenation, flow control, and buffering, acknowledgment and retransmission, priority handling, and header compression. The RLC layer can operate in different modes:

- *Transparent Mode (TM):* Used mainly for control information, with no retransmission or error correction.
- *Unacknowledged Mode (UM):* Used for real-time services where delays are critical (e.g., voice calls); it does not involve retransmissions.
- *Acknowledged Mode (AM):* Used for non-real-time services where reliability is more important; it includes mechanisms for acknowledgments and retransmissions.

The RLC layer encapsulates user data along with its headers before passing it to the MAC layer for transport over the radio interface.

9.9.2 NPSBN Data Service Design

9.9.2.1 PDN – Attachment Procedure

In NPSBN, the attachment or context method refers to the process by which a UE establishes a connection with the EPC to access network services. This involves several steps, including the initial signaling between the UE and the MME, the authentication of the UE, and the establishment of a bearer path for data transmission. This process ensures that the UE is properly identified, authenticated, and provided with the necessary resources for data connectivity.

Figure 9.9 describes a typical attachment call flow in NPSBN to get FN users connect the device with NPSBN network before actual transmission of the data packets.

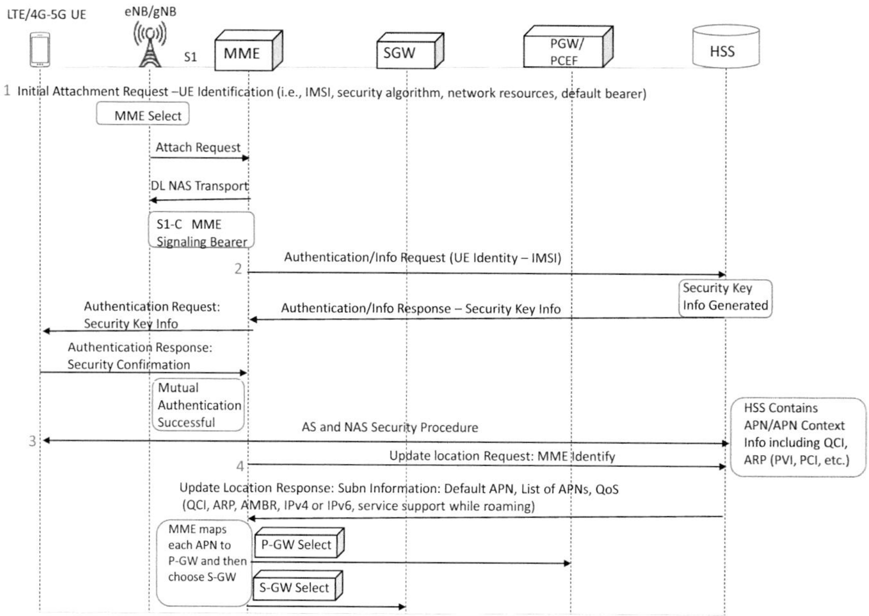

FIGURE 9.9
NPSBN UE Attachment Procedure.

The following steps briefly describe the actions during the attachment procedure:

1. Initial attachment request
2. UE authentication request
3. As and NAS security procedure
4. Update location request

At the end of the attachment procedure, the UE is identified, authenticated, and attached to SAE-GW in EPC.

9.9.3 Packet Data – EPS Bearer Setup Procedure

NPSBN EPS bearer tunnel or bearer tunnel is a data path established in LTE and 5G networks for transferring user data between a UE and the data network. It allows specific QoS parameters to be applied to different types of data traffic. The EPS bearer is a logical link that helps to manage and route the flow of data packets based on the service requirements and network policies.

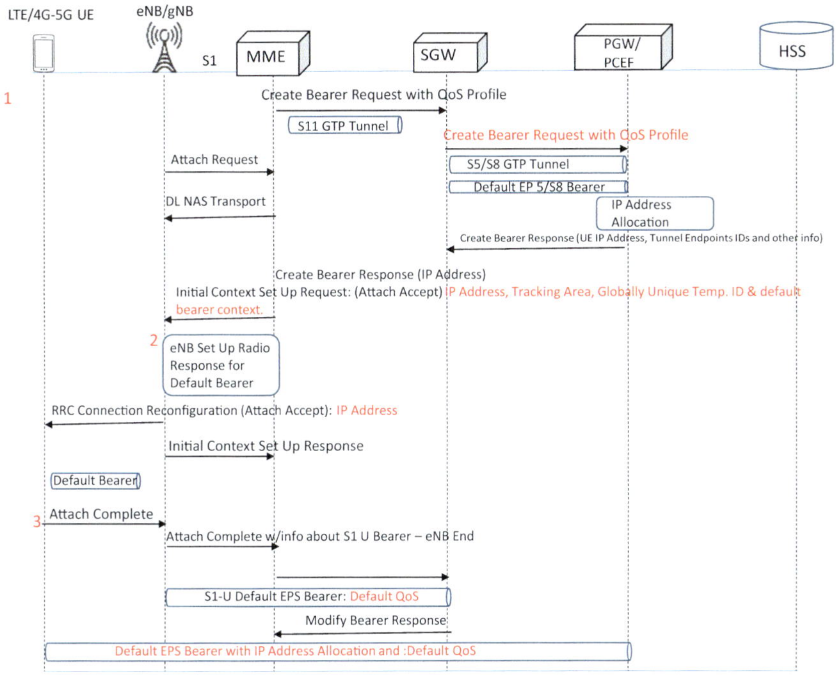

FIGURE 9.10
NPSBN Data Packet Tunnel Setup.

Figure 9.10 describes the various steps for the data bearer set:

1. Create bearer request with QoS profile
2. Prepare eNB/gNB setup for radio response
3. Attach SAE-GW to establish the default bearer

For the complete descriptions of the FN data QoS profiles for different FN user group, refer to Chapter 10, "NPSBN Quality of Service and Performance."

9.9.4 Data Service QPP and Performance

Data transmission across various FN users is one of the significant services of the NPSBN network enabling first responders to perform their tasks. However, it is also important that data transmission should be efficient, secure, and the quality of the data is not degraded when the device is trying to attach or during the data packet transmission over the NPSBN network.

TABLE 9.3

NPSBN Data Service QoS and DSCP

NPSBN Data Service	APN	QoS Class Indicator	DSCP	MPLS Exp
Mobile Data – High	firstnet_broadband	QCI 6	18	Default, EXP 2
Mobile Data – Normal	firstnet_broadband	QCI 8	22	Default, EXP 1

An FN device can be either in Idle Mode (IM) or in Dormant Mode (DM). The IM is when the device is not actively connected to the network, but it is still registered and can receive incoming calls or messages. The device monitors the network and can quickly reestablish a connection when needed. On the contrary, in DM, the device has a temporary suspension of its active connection, typically during data inactivity, but it can still receive paging messages from the network.

NPSBN has fully incorporated 3GPP specifications – 3GPP TR 36.913 (41) and 3GPP TR 25.913 (42). The target for transition time from IM (with IP address allocated) to DM is > 50 ms for Control Plane (S1 C) latency. The User Plane (S1 U) specifications for DL is 300 Mbps and for UL is 75 Mbps. In 5G, the target for control plane latency is significantly lower than in previous generations, aiming for around 10 milliseconds or less in optimal conditions.

Table 9.3 provides a summary of the data service QoS and Differentiated Services Code Points (DSCP).

Mobile high priority data in LTE refers to data traffic that is given precedence over normal priority data, ensuring faster and more reliable delivery, especially during network congestion. This prioritization is particularly beneficial for applications requiring low latency or high reliability. Normal priority data does not receive the same level of prioritization, which can lead to slower transmission rates under crowded network conditions. It should be noted that these APN, QoS Class Identifier (QCI), DSCP, and Multiprotocol Label Switching Expansion (MPLSS Exp) are same for 4G and 5G for data services.

9.9.5 Data Service – Key Points

NPSBN data service enables faster, reliable, and secure data transmission between UE and PDN or internet access for smartphones, tablets, and other mobile devices, enhancing the user experience for streaming, browsing, and downloading and uploading content.

NPSBN data services use FN QPP, which provides prioritization of access and QoS for priority data and the ability of offloading non-FirstNet users during congestion.

9.10 REAL-TIME TESTING

Real-Time Testing (RTT) is an alternative to TTY for customers who are hearing or speech impaired. TTY devices allow users to send typed messages over telephone lines. However, the prevalence of TTY devices has declined with the rise of more advanced wireless communication technologies, e.g., 4G/5G networks. The Federal Communications Commission (FCC) has suspended the TTY requirements for emergency services due to the rise of new technologies. It has amended its rules via FCC 16-169 (43) and RFC 4103 (44) to facilitate transition from text TTY technology to RTT as a reliable and interoperable universal text solution over wireless IP-enabled networks for people who are deaf, hard of hearing, deaf-blind, or have a speech disability (collectively, "people with disabilities" or "text-reliant users").

9.10.1 RTT Service Architecture

RTT sends text immediately as it is typed. It is especially useful for making regular or emergency calls. RTT is different from instant messaging and SMS. The user starts a text conversation like a voice call and both parties see text characters appear on their devices as they are typed. RTT service is available to FirstNet users as a native dialer integrated solution and supported by FN-enabled devices having FirstNet SIM. Figure 9.11 shows the TTY devices that are getting replaced with LTE/4G devices, enabling enhanced services including RTT.

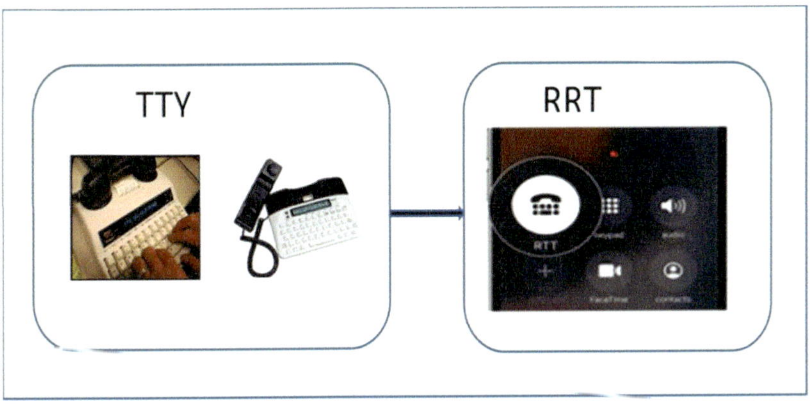

FIGURE 9.11
Older TTY and NPSBN RTT Devices.

9.10.3 RTT Service Architecture and Design

RTT service architecture typically involves several key components, including:

1. *RTT Applications:* Mobile apps or web interfaces where users send and receive RTT.
2. *WebSocket Server:* Manages real-time communication between clients, enabling instant message delivery.
3. *Application Server (AS):* Handles business logic, user authentication, and message processing.
4. *Database:* Stores user profiles, message history, and other relevant data.
5. *Notification Service:* Sends push notifications to users for new messages when they are not actively using the app.
6. *API Gateway:* Manages API requests and routes them to the appropriate services.

Figure 9.12 illustrates a design architecture of NPSBN RTT service available to FirstNet user having a 4G/5G-enabled device with FN SIMs. This RTT architecture enables efficient, scalable, and real-time communication between users and emergency 9-1-1 dispatchers.

9.10.4 RTT Service QPP and Performance

Table 9.4 provides the NPSBN RTT service QPP. The QPP and performance are dependent on the FN UE's mode of access to RTT service, i.e., over eNB/gNB or over Wi-Fi. Session Boarder Controller (SBC)/ Proxy Call Session Control Function (P-CSCF) establishes a separate RTT bearer over the cellular access network using IMS APN. If the network access is over Wi-Fi, SOS APN is used and the best effort QPP and performance are applicable.

TABLE 9.4

NPSBN RTT Service QPP and Performance

Radio Access Technology	Voice Bearer	Text Bearer	APN
Device on eNB/gNB	QCI 1, GBR	1.7 Kbps, Non-GBR, QCI-8, PCI=0, PVI=1	IMS APN (RTT)
Device on Wi-Fi	Best-effort VoIP	Best efforts QPP	SOS APN (RTT)

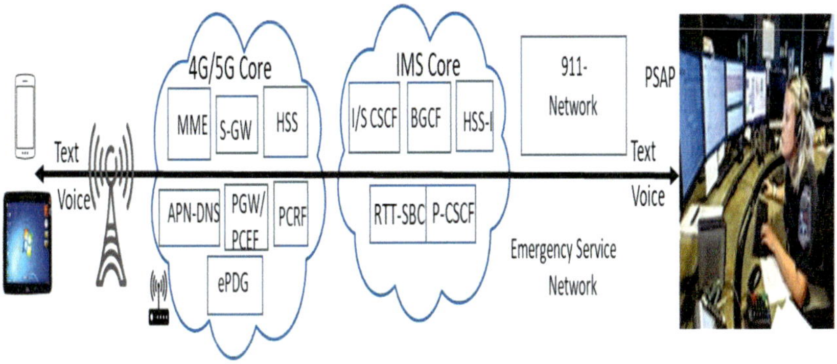

FIGURE 9.12
RTT Service Design Architecture and Design.

9.10.5 RTT Service – Key Points

- FN RTT can be accessed by FN users on NPSBN RAN as well as on Wi-Fi using FN-enabled devices (FN 313-100 PLMN) available to first responders.
- Launching of RTT from the device is through a dedicated button for RTT on the device. RTT is natively integrated with NPSBN core networks.
- Simultaneous voice and text communication are enabled. RTT-capable devices send and receive both text and voice simultaneously in both directions over IP on the same call session.
- RTT call for e9-1-1 is routed to local PSAP where 9-1-1 call is typically answered by emergency dispatch operators or call takers who are trained to assess the situation and dispatch necessary services like police, fire, or medical help.
- RTT service is a regulatory requirements (e.g. 7-1-1 and 9-1-1). Therefore, the RTT interworking function is included in the design to ensure that VoLTE RTT calls on 4G/5G RAN or Wi-Fi can connect to TTY users on the PSTN, including 7-1-1 and 9-1-1 operators, as per 3GPP specifications TS 22.226 (45).

9.11 NUMBERSYNC (TWINNING) SERVICE

NPSBN users generally use mobile smartphones and mobile smartwatches, also called wearable. FirstNet NumberSync service is a feature that is used to "twin" a smartphone with smartwatch to enable the watch to make and receive phone calls and send/receive text messages which are received over the smartphone number. In NumberSync service, the smartphone is also referred to as Primary Device (PD) and the smartwatch as Secondary Device (SD). The presence of the PD in the vicinity of the SD is not essential for smartwatches to communicate.

Number synching of the smartphones and wearable devices enhances first responders experience as they can use either of the devices for VoLTE calls and send/receive text messages. Also, the user can use the wearable in place of their FN-enabled smartphone in case phone is not with them or situation does not warrant to use the phone.

Relevant specifications can typically be found in the 3GPP TS 23.204 (46) series, which addresses various aspects of device synchronization. 3GPP TS 23.057 (47) deals with the synchronization of mobile devices in general. 3GPP TS 24.229 deals with the SIP signaling and includes aspects related to number synchronization in voice and services.

9.11.1 NumberSync for Apple and Non-Apple Devices

FN supports a large collection of FN devices of various makes, models, and features. First responders can select the device for their day-to-day activities. NPSDB NumberSync service has two distinct solutions depending upon the make of the devices.

1. NumberSync for Apple IOS-based devices: This solution is basically for the Apple phone and Apple smart wearables due to distinct security and messaging features in Apple IOS. Currently, Apple wearable can work with iPhones since Apple IOS has built-in authentications and messaging features. All iPhones and Apple watches are based on Apple IOS-based NumberSync solution.
2. NumberSync for non-Apple IOS devices which are based on Android, and Tizen IOSs: For example, Samsung, Google, and LG devices are based on Android. Samsung uses Tizen IOS for their watches made before their Galaxy watch series. Samsung Galaxy, Google Pixel, and other Android-based phones and watches are based on this NumberSync solution.

9.11.2 NumberSync Service Architecture

Figure 9.13 provides a high-level NumberSync service architecture presenting Android-based devices. As stated earlier, NumberSync solutions for Apple

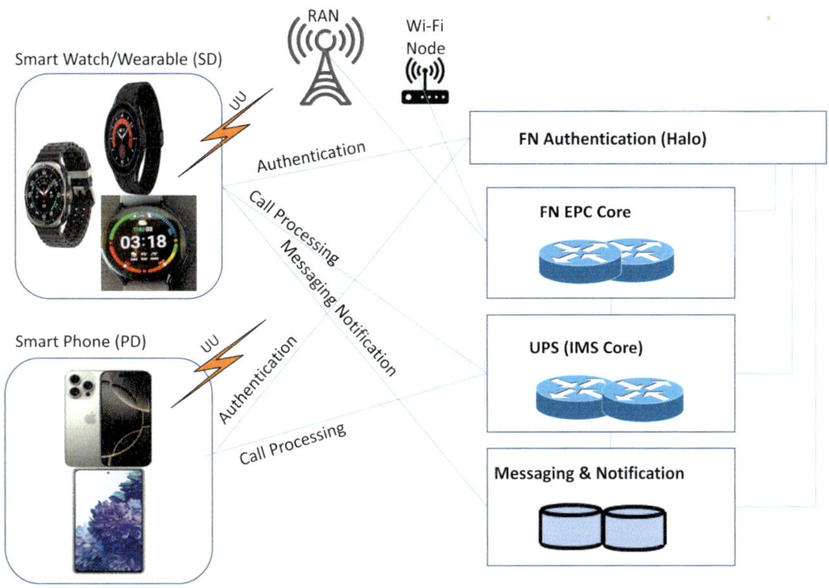

FIGURE 9.13
NPSBN NumberSync High-Level Logical Architecture for Android iOS Devices.

devices are somewhat less complicated compared to Android IOS-based NumberSync. Establishing "twinning" relationship is required between a PD and an SD for NumberSync service to work.

Both PD and SD complete the initial attachment process with RAN once their power is on. The twinning process is initiated via an application loaded on PD. PD and SD are authenticated by the FN authentication system "Halo." After twinning authentication, messaging authentication is completed and connectivity is established with the IMS core and the messaging and notification systems. At the completion of process, PD and SD are twinned enabling SD to make and receive phone calls and send/receive text messages.

9.11.3 NumberSync Service Design and Call Flow

The following are the key design features of the FN NumberSync service.

1. *Twinning:* The NumberSync relationship is established using a "twinning" process that is launched only by the twinning application on the PD. PD is always a smartphone while the SD is always a watch/wearable.

2. *NumberSync Core Support:* A FirstNet user launches NumberSync when both PD and SD devices are on the FirstNet core (313-100).
3. *Service Order Code (SOC):* A specific SOC is required for NPSBN NumberSync service.
4. *Cross-BAN Support:* The feature is supported when one device is on agency paid account (PLMN 313-100) and the other is non-agency paid (e.g., FN individual user's account – PLMN 313-100 – or an AT&T consumer account – PLAMN 310-410).
5. *Device Compatibility:* NumberSync service will not facilitate twinning between two smartphones or between two wearables. Twinning is only achieved between a PD and an SD.
6. *QoS Service Levels:* The QoS service varies depending upon class of users:
   - Primary users have options for platinum, gold, or silver levels of services. Initially, these users are assigned silver level as default but can upgrade as desired.
   - Extended primary users are initially assigned the default level which is upgradable to silver.
   - Individual users have only the default level.

9.11.4 FN NumberSync versus Bluetooth Features

NumberSync and Bluetooth are two different service features supported by FN wireless devices. Bluetooth devices use short-range wireless technology (frequencies in the 2.4 GHz ISM band) to exchange data over short distances typically >300 ft. between a PD (e.g., mobile phone) and the Bluetooth device. A connection is established between two devices (PD and Bluetooth device) through a process called pairing, which involves discovering the available devices, establishing a secure link, and then transferring data in small packets over radio waves. Bluetooth devices are not directly connected to the network but only paired with the PD and must be in close proximity to get Bluetooth connectivity.

On the contrary, in the NumberSync service, both PD and wearable device (SD) are connected with 4G/5G wireless or over a Wi-Fi network and have separately assigned ten-digit North American Numbering Plan (NANP) – three-digit area code followed by a seven-digit local number, formatted as (NXX)-NXXXXXX. Once the PD and SD are twinned, sharing of data packets begins between PD and SD. Here are the key points of NumberSync services.

- Both PD and SD require separate ten-digit phone numbers.
- An SD can be twinned with only one PD at a time, but a PD can be twinned with a maximum of six SDs. Twinning is done by an App on PD.
- Once twinned, SD acts like a PD, makes/receives calls, and exchanges data packets over networks.
- It is not required for PD and SD to be in proximity or nearby. If they are connected to a network, PD can be left at home, office, or any other locations, yet SD will work.
- If SD is not twinned, both PD and SD will work independent of each other and information will not be shared between them.

9.11.5 FN NumberSync Security

NPSBN is required to adhere to identity, credential, and access management. For the NumberSync service, rightful users of the PD and SD must be authenticated by FN security systems as part of the twinning process. Also, messaging and notification systems will authenticate SD for sending the notifications and sharing the messages.

9.11.6 NumberSync Service Performance and QPP

NPSBN NumberSync service performance and QPP are entirely dependent upon the FN user profile and is not governed by PD and SD vendors. In addition, FN QPP is equally applicable to FN users with Corporate Responsible User (CRU) and Individual Responsible User (IRU). CRU and IRU defer in the ownership of the devices, whether the devices are owned and registered by a PSE or by an individual first responder.

9.11.7 NumberSync Service Key Points

NPSBN NumberSync is an important service to improve work experience of first responders. The following are the key points:

- NumberSync service enables first responders to use either smartphones (PD) or wearable devices (SD) to make VoLTE calls, texting, or data exchange.
- PD and SD need not be in proximity to each other since both devices are attached to NPSBN.

- Both PD and SD have separate ten-digit phone numbers. The PD and SD act as the same unit if twinned. If not twinned, PD and SD work as two separate units.
- PD and SD are authenticated by NPSBN and meet NPSBN security requirements.
- PD and SD have FN QPP and meet NPSBN performance requirements.

9.12 OUTBOUND ROAMING SERVICE

FN outbound roaming service enables the first responders or FN subscribers to use their FN-enabled devices to access services when traveling outside the NPSBN network's coverage area, especially attaching the networks of other service operators. This process involves the home network (FN subscribers' original network, i.e., NPSBN) allowing the user to connect to a visited network and encompasses services like voice, SMS, and data. The visited networks include roaming service providers within the United States as well as any international locations.

In LTE, outbound roaming is also called "S8HR" since it uses the S8 interface, which is part of the 3GPP specifications for LTE networks (48). S8 interface connects the NPSBN SGW in the home network (i.e., NPSBN EPC core) to the SGW in the visited network, facilitating data transfer while maintaining a secure connection. The "HR" stands for "Home Routing," indicating that the data traffic is routed back to the home network for handling. This is called the S8HR (S8 Home Routed) architecture. IMS functionality resides only in the Home PLMN (HPLMN) (49). IMS APN is determined to connect to the nearest PGW in the HPLMN. The creation of the S8HR architecture removes the dependency of the HPMN and Visited PMN (VPLMN), which deploy IMS and simplify the roaming interface as well data security.

FN/AT&T have roaming agreements in place with the roaming service providers, also called roaming partners, enabling FN users to roam on their network while in their coverage areas. It is also important to understand that the quality and availability of roaming services can vary by the visited network and roaming agreements. FN/AT&T does not allow inbound roaming on NPSBN, i.e., non-FN subscribers cannot roam/access NPSBN. An outbound S8HR VoLTE roamer on a partner LTE network will be treated in much the same way as a VoLTE user on the NPSBN from the IMS perspective.

The following definitions apply for better understanding of the outbound roaming:

- *S8HR:* Refers to the use of the S8 GTP interface (SGW to PGW in EPC) to support the IMS APN.
- *HPLMN:* Refers to the home network provider of the FN VoLTE/IMS roamer. NPSBN will be the HPLMN provider when its subscriber is roaming outbound on a visited partner LTE network.
- *VPLMN:* Refers to the visited LTE network serving the FN VoLTE/IMS roamer. VPLMN always denotes to roaming partner's 4G/5G network.

9.12.1 FN Outbound Roaming Service Architecture

The FN S8HR service enables FN users to make or receive VoLTE calls while roaming on a partner carrier's LTE RAN by establishing a connection between the partner carrier's Packet Core and NPSBN EPC.

Figure 9.14 depicts interfaces between HPLMN and VPLMN networks. FN/AT&T's IXC zone uses a firewall to interconnect external network service providers including the roaming partners. Also, the roaming connectivity between two interconnecting partners is facilitated via a third party IPX service provider. IPX, or IP Exchange, refers to a service provider that

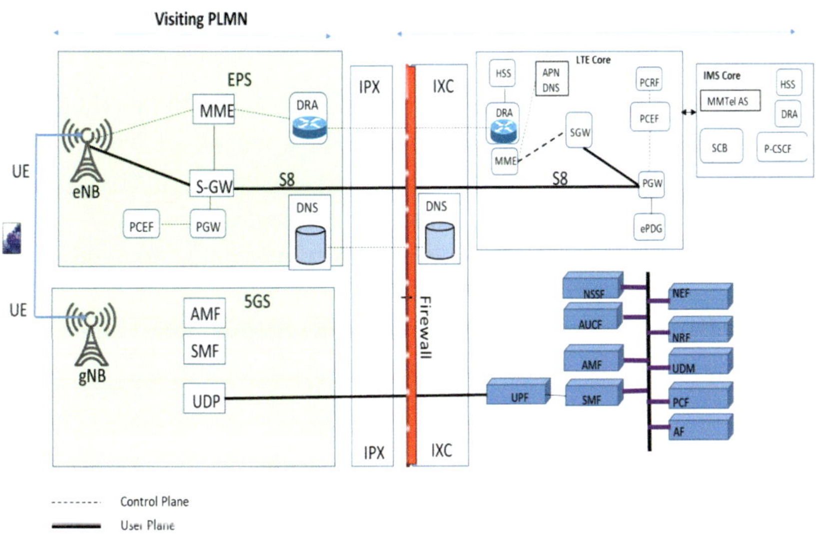

FIGURE 9.14
S8HR Outbound Roaming Logical Architecture.

facilitates the exchange of IP-based services, such as VoIP and mobile data services, among different networks. IPX providers enable secure, high-quality communications by connecting various mobile operators and service providers within a global framework, allowing seamless interoperability and traffic exchange. NPSBN uses IP exchange service provider, Syniverse, which connects NPSBN to roaming partners to provide roaming services.

Figure 9.14 also illustrates that first responders having 4G/5G-enabled devices can roam on any NPSBN roaming partner's 4G or 5G networks having coverage in domestic or global markets.

9.12.2 FN Outbound S8HR Roaming Design and Call Flow

NPSBN outbound S8HR roaming service involves significantly complex service design. The complexities in an outbound LTE roaming network include various technical and operational challenges that arise when a first responder travels outside the NPSBN network's coverage area and connects to a roaming partner's network whether within the United States or outside the United States. These include issues related to network interoperability, signaling management, billing processes, and ensuring seamless service quality. In addition, complexities involve the integration of different network technologies, regulatory compliance across different jurisdictions, and handling of different UE capabilities.

In the NPSBN S8HR outbound roaming, the signaling and media use same IMS APN. This design has several advantages:

1. Provides significantly faster connection and call setup enabling low latency.
2. Provides RCS, an advanced messaging protocol in LTE services designed to enhance the standard SMS messaging experience providing rich features that go far beyond simple text messaging, like voice, video, file transfer, group chat.
3. Provides enhanced security by creating an IPsec tunnel for SIP signaling between UE and SBC enabling
   - 3.1. Data confidentiality
   - 3.2. Data integrity authentication
   - 3.3. Data origin authentication

Figure 9.15 describes briefly a typical S8HR outbound roaming call flows for initial FN UE attachment to roaming partner's Ran/MME and subsequently

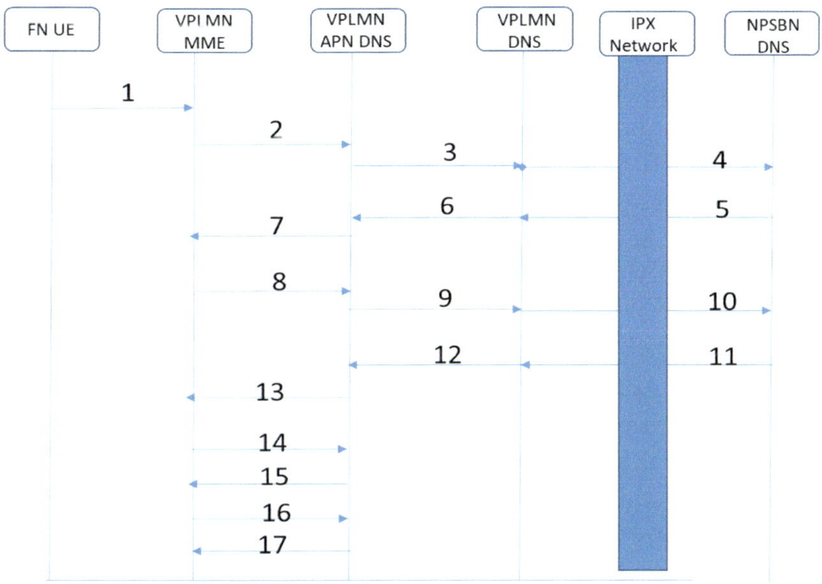

FIGURE 9.15
NPSBN S8HR Outbound Roaming Call Flows.

S/PGW selection. As stated earlier, the data packet in S8HR roaming is exchanged between VPLMN SGW and HPLMN PGW. In addition, Naming Authority Prefix Translation Resolver (NAPTR) is invariably used for service discovery and resolution. It enables the translation of domain names into Uniform Resource Identifiers (URIs) and manages various service types such as voice, video, or messaging over IP networks. NAPTR records are part of the DNS and are used to discover services and facilitate communication sessions based on the user's preferences or requirements. APN's "ims" and "FirstNet_phone" are configured in IXC DNS to enable NPSBN outbound roaming.

The flows are listed here:

1. FN UE requests bearer setup from VPLMN MME with configured NPSBN HPLMN VoLTE APN (Firstname_phone).
2. VPLMN MME issues S8 Naming Authority Pointer (NAPTR) query to VPLMN APN DNS for NPSBN HPLMN VoLTE APN.
3. VPLMN APN DNS searches for an existing record. If no record is found, it forwards recursive S8 NAPTR query to VPLMN IPX auth DNS.
4. VPLMN IPX auth DNS performs a search for the NAPTR record. If no record is found, VPLMN IPX auth DNS performs iterative query via a

server-selected IPX partner network and is referred to NPSBN/AT&T IXC DNS.

5. NPSBN/AT&T IXC DNS issues a response for the requested VoLTE APN based on source IP of VPLMN IPX DNS
6. VPLMN IPX DNS receives VoLTE APN NAPTR response and sends response back to VPLMN APN DNS.
7. VPLMN APN DNS looks in cache of VoLTE APN NAPTR response from VPLMN IPX DNS and forwards it to VPLMN MME.
8. VPLMN MME selects a PGW from the NAPTR 'set' sent and sends a PGW record request to VPLMN APN DNS.
9. VPLMN APN DNS examines its current cache. If no record is found, it forwards recursive S8 PGW record query to VPLMN IPX DNS.
10. VPLMN IPX DNS looks in cache for requested PGW record. If no record is found, IXC DNS performs recursive query via server-selected IPX partner network and is referred to ATT IXC DNS.
11. ATT IXC DNS sends authoritative response for requested PGW record to VPLMN IPX DNS.
12. VPLMN IPX DNS caches PGW record response and sends response back to VPLMN APN DNS.
13. VPLMN APN DNS caches PGW record response from VPLMN IPX DNS and forwards it to VPLMN MME.
14. VPLMN MME sends S8 NAPTR request for VoLTE roaming SGW selection to VPLMN APN DNS.
15. VPLMN APN DNS responds with S8 VoLTE SGW NAPTR set based on TAC HB code in query.
16. VPLMN MME selects S8 SGW from returned NAPTR set and issues a record query to VPLMN APN DNS for SGW hostname.
17. VPLMN APN DNS replies with a record response for requested S8 SGW hostname.

NPSBN CTS-generated Call Detail Record (CDR) contains detailed information about the calls, such as call duration, time of the call, and the parties involved, which are essential for billing purposes. The NetLoc procedure defined in 3GPP TS 29.212 (50) and 29.214 (51) is used to request and deliver the cell site data or Access Network Information (ANI) to the NPSBN IMS core which is recorded in the FN CTS CDRs. The ANI includes the current cell site identification known by the network for the user as well as time zone information if it is available.

9.12.3 NPSBN Outbound Roaming QPP and Performance

The performance of NPSBN VoLTE outbound roaming service will be dependent on the QPP availability and reliability of the VPLMN LTE/5G RAN and EPC. FN/AT&T executes roaming contracts with each roaming partner which includes negotiation of the FN roaming QPP. However, the VPLMN portion of the network is outside of FN control. NPSBN always assures the acceptable latency and reliability in the portions of the NPSBN network under its control and management.

NPSBN supports S8HR outbound setting of the IMS APN having QCI= 1 for voice, QCI=6 for data, and QCI=5 for IMS signaling. The end-to-end reliability requirements for IPX links and IXC zone is 99.999% availability. The IXC zone plays a critical role in NPSBN roaming, particularly ensuring secure and efficient interconnection between NPSBN/AT&T and the partners' telecommunications networks.

NPSBN/AT&T has developed a set of procedures and probes to monitor the roaming operational health. Outbound roamer data session traces are collected and analyzed to pinpoint the issues and correct promptly.

9.12.4 FN Outbound Roaming Partners

U.S. Cellular is a major S8HR roaming partner in the Mainland U.S. and Roger Communications in Canada. FN/AT&T also has roaming partners in Puerto Rico, U.S. Virgin Islands, Guam, CNMI, and American Samoa.

9.12.5 FN Outbound Roaming – Key Points

- Provide FirstNet subscribers an outbound VoLTE/IMS roaming service using FN SIMs (PLMN 313-100) for both domestic and international roaming.
- FN roaming UE is encouraged to roam to a preferred network to roam indicated by the HPLMN using Steering of Roaming (SOR) technique.
- No inbound roaming is allowed on NPSBN network.
- FN IMS APN is resolved to a PGW in the HPLMN reducing complexity and the connection time.
- Conventional PSTN dialing services, e.g., 800, 9-1-1
- VoIP over Wi-Fi while outbound roaming

9.13 NPSBN MISSION CRITICAL SERVICES (MCPTT, MC VOICE, MC DATA, MC VIDEO)

The 3GPP Mission Critical Service Technical Specifications TS 22.280 (52), TS 22.179 (53), TS 22.281 (54), and TS 22.282 (55) outline the requirements and functionalities needed to support emergency and Mission Critical Services, collectively called MCX, in NPSBN. It includes specifications for real-time voice, data, and video services and group communications tailored for first responders and other emergency services. The framework facilitates reliability and priority access during critical emergency situations.

Table 9.5 provides the 3GPP evolution of MCX covering both 4G and 5G networks.

The standards have categorized the following group of services which are also termed as MCX and provided by NPSBN for FN users.

1. Mission Critical Voice (MC Voice) Including Mission Critical Push-to-Talk (MCPTT)
2. Mission Critical Video (MC Video)
3. Mission Critical Data (MC Data)

On the basis of the nature of the emergency and the scope, a first responder could be handling an emergency alone as a member of an incident management team or as a group. As an individual, for example, first responder

TABLE 9.5

3GPP Evolution of MCX for 4G and 5G Networks

3 GPP- Evolution of Mission Critical Services for 4G and 5G		
Release 13 (March 2016)	**Release 14 (June 2017)**	**Release 15 (Nov 2018) & Release 16 (March 2020)**
• Mission Critical Push To Talk over LTE (MCPTT)	• Mission Critical Video (MCVideo) • Mission Critical Data (MCData) • Mission Critical Service Common Requirements (MCCoRe) • Mission Critical Push To Talk Ove LTE Re-alignment (MCPT-R) • Enhancements of Previous Features (Group)	• Enhancements of MCVideo, MCData, PCPTT and MCCORE • 5G Support for Quality of Service (QoS), Priority and Preemption • Support for Dynamic Controller • Improvement of Direct Mobile Capabilities

generally receives priority dispatch call from e9-1-1 command center for a motor vehicle accident the accident location information and vehicle descriptions to attend the emergency. In other cases, a first responder may be part of a group communication handling a much larger emergency, e.g., explosions, fires, terrorist attacks, which may impact many lives and properties.

NPSBN supports MCX services in the following manners depending upon the number counts of the participants:

1. *Unicast or Private Cast:* Unicast MCX involves one-to-one connection between two first responders, allowing a single first responder to communicate directly to another first responder. This method is typically used for private conversations and exchange of critical information.
2. *Multicast or Group Cast:* Multicast MCX communication allows a single first responder to communicate to several first responders simultaneously. This is useful for group communications where the sender wishes to communicate with several recipients at once without needing to establish individual connections. The MCX service administrator or incident command may restrict who can be a member of specific MCX service groups, so that those MCX service groups shall be inaccessible to other users, including dispatchers or supervisors. An individual can be a member of more than one group simultaneously.

NPSBN supports MCX services in the following manners depending upon the nature, severity, and the immediate needs of the situations.

1. **Emergency call**:
   a. **Group call** Communication involving an incident where the first responder's life is in immediate danger. The network assigns bearer resources with QCI 65 and Allocation and Retention Priority (ARP) 2.
   b. **Private emergency call** requires the admin to provision an emergency group between two users. The network assigns bearer resources with QCI 65 and ARP 2.
2. **Imminent peril call:** Group call communication involving an incident where the first responder is dealing with a life-threatening situation. Only an authorized dispatcher can invoke this call type from the console and the network assigns bearer resources with QCI 65 and ARP 2.

3. **Broadcast call**: A high-priority announcement from the command center, e.g., dispatcher. The network assigns bearer resources with QCI 65 and ARP 3, ensuring the broadcast is delivered to the group with a higher priority.
4. **MCPTT normal priority call**: Group call communication for a group that is configured with normal priority by the agency admin. The network assigns bearer resources with QCI 65 and ARP 3. If required, the incident commander/dispatcher can dynamically uplift the priority of this group as an emergency call.
5. **Private nonemergency call**: Normal private call communication between two MCPTT users. The network assigns bearer resources with QCI 65 and ARP 5.
6. **Ad hoc group:** Created by a user from the same prearranged groups or across different prearranged groups. The network assigns bearer resources with QCI 65 and ARP 3. This group will disappear from the list after the user logs out because no server update occurs with this information.

Not all MCX services on a device get emergency priority. When an FN user initiates an emergency group communication, only the active MCX services are granted emergency priority based upon the prevailing situations. In case of "Imminent Peril," only the active MCX service receives the imminent peril priority.

Multimedia Broadcast Multicast Service

3GPP (56) is constantly enhancing the MCX requirements to add new features and service support to first responders. Multimedia Broadcast Multicast Service (MBMS) is a technology used in 4G and 5G networks to efficiently deliver multimedia content, such as video and audio, to multiple users simultaneously. This service allows NPSBN to broadcast content over the network, reducing the bandwidth required for delivering the same content to individual users.

MBMS "point-to-multipoint" technology enables the same multimedia content delivered simultaneously to multiple first responders. This "one-to-many" distribution mode can be a valuable alternative to unicast in supporting "large group calls" for a large number of first responders interested in the same content. MBMS uses the inherent broadcast qualities of NPSBN to send the voice/video or data only once to reach several first responders, allowing efficient use of radio access and transport resources. In 4G, MBMS enables operators to provide services like mobile TV and live event broadcasting, optimizing resource usage by sending a single stream of data to multiple users. In 5G, MBMS is enhanced with features such as improved scalability and lower latency, allowing richer multimedia experiences and more efficient use of network resources.

NPSBN MCX Application Client

NPSBN MCX supports both integrated clients and the downloadable clients complying with 3GPP TS 23.280 (57) to enable a variety of FN devices for MCX. A preloaded client is integrated with the device Operating System (OS) having a dedicated MCPTT APN. When the MCX client starts, the device opens a default bearer automatically with APN "mcptt" without an attachment request.

An MCX client is downloaded to the device like other applications and is configured with the general-purpose APN "firstnet_phone" and establishes a default bearer with QCI 6. When the MCPTT client starts the MCPTX AS, an Rx request is sent to establish a dedicated QCI 65 bearer.

9.13.1 MCX Architecture

MCX architecture components based on 3GPP TS 23.280 are described here:

1. *The Common Services Core (CSC) Server:* includes the Configuration Management Server (CMS), Group Management Server (GMS), and Key Management Server (KMS).
2. *The Identity Management Server (IdMS):* Uses ICAM and HALO platforms.

   The two basic types of identities include Application Layer ID and Signaling Layer ID. Both MC ID and MCPTT ID are stored in the NPSBN HALO IdMS platform.
3. *AS:* performs Group Communication Services Application Server (GCS AS) for supporting MCX services.

Figure 9.16 illustrates the high-level architecture of the NPSBN MCX.

It also shows the interactions among MCX components including FN MCX-enabled UE, RAN (eNB/gNB), cell site backhaul/transport, and packet core. The component includes an MCX SBC complying with standards specifications for MCX and MCX System and AS. These components authenticate devices and device users to access the NPSBN MCX and perform desired functions (e.g., MCPTT, MC Video, MC Data) for unicast or multicast situations. MCX and applications are continuously evolving and first

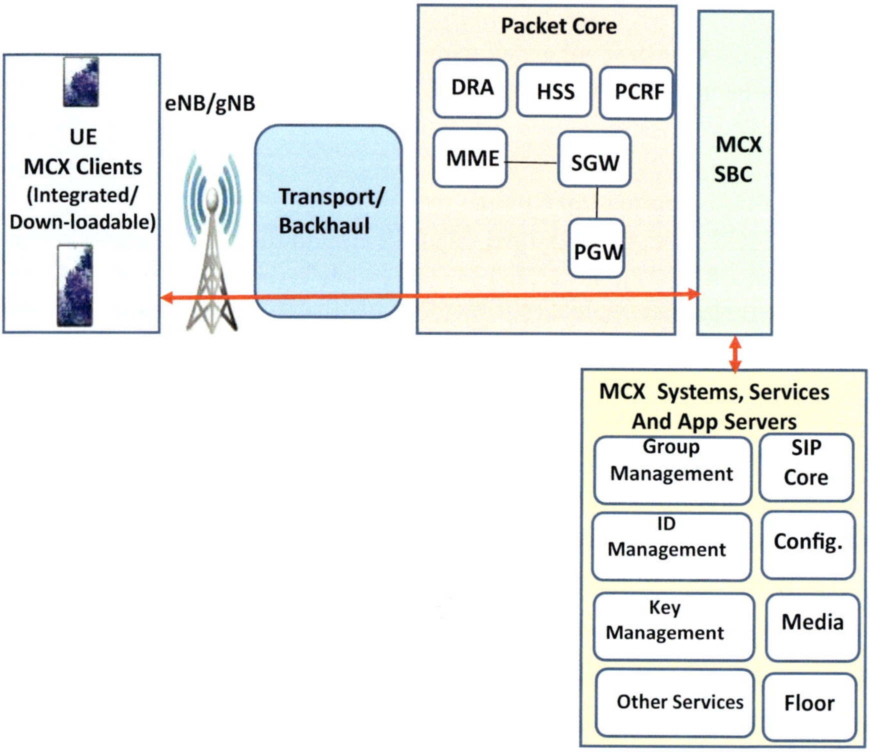

FIGURE 9.16
NPSBN MCX – Logical Component Architecture.

responders will find more robust and newer MCX supported by additional varieties and make of UEs.

KMS functional entity stores and provides security-related information (e.g., encryption keys) to the key management client, group management server, and MCX server(s) to ensure confidentiality and integrity of media and signaling.

SBC supports all the traffic flow generated from any endpoint interacting with MCX system. The SBC platform is dedicated to MCX and securely passes MCPTT client traffic to MCX AS. The client traffic includes SIP, HTTPs, XCAP, MBCP, and media using secure Real-time Transport Protocol/ Real-time Transport Control Protocol (RTP/RTCP) and supports IPV4–IPV6 source.

9.13.2 MCX Bearer Design

The MCPTT EPS bearer's initial attachment procedures may use a dedicated mcptt APN or the general-purpose firstnet_phone APN depending on the client type.

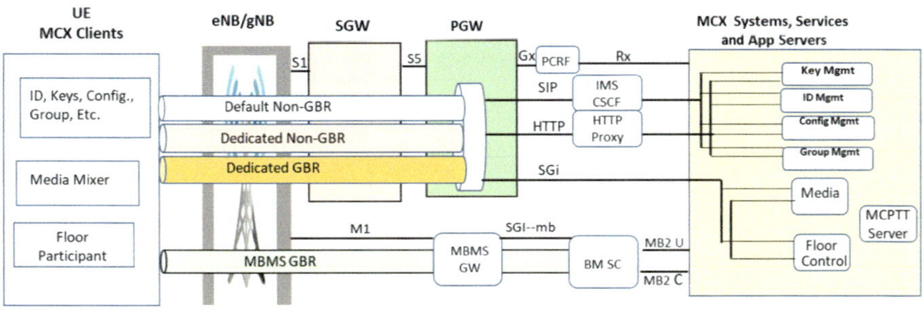

FIGURE 9.17
MCX – Bearer Design.

Figure 9.17 illustrates the MCPTT-integrated client EPS bearers from the UE to the PGW using a dedicated mcptt APN. The default bearer QCI 69 attributes are assigned during the initial PDN connection procedures. The media bearer uses an Rx request from the MCPTT application to the PCRF to establish or remove depending on the call activity. The external interface SGi illustrates the MCPTT services delivered to the MCX servers where data services are shown generically flowing to the internet.

Also, a Guaranteed Bit Rate (GBR) tunnel is used for MBMS between UE and BM SC. FN user can select "Media Mixer" and "Floor Participant" buttons to configure and manage multicast.

9.13.3 MCX QPP and Performance

NPSBN provides first responders with several services including MCX with enhanced QoS and upmost performance is expected in emergency situations.

3GPP technical specifications (58) (59) dealing with MCX to support public safety at large include new QCIs for mission critical public safety applications which are currently implemented into NPSBN. GBR QCI 65 and QCI 66 are for mission critical PTT voice and non-mission critical PTT voice, respectively. Non-GBR QCI 69 is for mission critical signaling and non-GBR QCI 70 is for mission critical data. Non-GBR and GBR QCI bearers are defined by 3GPP.

GBR ensures that bits delivered per second are met, whereas non-GBR is not guaranteed. QCI is a scalar value that controls packet forwarding treatment in the 4G/5G access network. The standardized QCI characteristics for the bearer traffic between the device and the gateway are in terms of bearer types: GBR or non-GBR, priority, packet delay budget, bit rate, and packet error loss rate. ARP is based on three pieces of information: priority level, preemption capability,

TABLE 9.6

Mission Critical QCI and Performance Data

QCI	Bearer	Packet Delay (ms)	Packet Loss	Use Case
65	GBR	75	10^{-2}	MCPTT User Plane
66		100	10^{-2}	Non-Mission Critical User Plane
69	Non GBR	60	10^{-6}	Mission Critical delay sensitive signaling
70		200	10^{-6}	Mission Critical Data

TABLE 9.7

NPSBN MCX Call Types and QPP

	Dedicated Bearer	
MCPTT Call Types	**QCI**	**ARP**
Emergency Group	65	2
Emergency Private	65	2
Imminent Peril Group	65	2
Broadcast Group	65	3
Normal Group	65	3
Ad hoc Group	65	3
Normal Private	65	3

and preemption vulnerability. ARP determines whether bearer creation/modification requests are accepted or refused when resources are constrained

Table 9.6 describes the recommended QCI and other relevant performance information for MCX.

Table 9.7 shows QPP allocation based on the call types described in this chapter earlier. Initial allocation of the QPP is dependent on the MCPTT clients supported on the devices. An MCPTT-integrated client will always use dedicated APN "mcptt," enabling the first responders to connect dedicated bearer and QCI 65. The first responder using the OTT client will initially get default bearer and QCI 69 assigned. However, the moment they start the MCPTT call, they will always be uplifted to QCI 65.

9.13.4 MCX – Key Points

MCX are constantly evolving to enhance first responder abilities to share information during emergency handle and better user experience. Here are the key features:

- NPSBN fully complies with 3GPP and other standards in architecting and developing MCX.
- MCX including MBMS enable group communications, which results in saving valuable resources.
- MCX QCIs are dependent on the roaming partner's network capabilities during roaming.
- MCX architecture is enabled to older communications services, e.g., LMR which are still in use in many public safety agencies.
- NPSBN MCX are engineered with geo-redundancy and 24/7/365 monitoring to guard against physical and cyber threats.
- NPSBN MCX provide web-based administration, enabling dynamically proved talk group functions, and share interagency resources for near real-time cross-collaboration.

9.14 EMERGENCY E9-1-1 SERVICE

Emergency e9-1-1 service enables 9-1-1 callers to request immediate assistance from police, fire, or medical services. When someone dials 9-1-1, they are connected to a Public Safety Answering Point (PSAP). A PSAP is a call center responsible for answering emergency calls, such as those made to 9-1-1. A call to 9-1-1 is routed to the nearest PSAP, where trained dispatchers/ call takers answer the call, gather essential information, assess the situation, and send the appropriate emergency services (police, fire department, or ambulance). NPSBN emergency service enables first responders to make emergency calls. When first responders make a 9-1-1 call, they typically do so to request emergency assistance, report an incident, or notify dispatcher about a situation requiring additional resources. This process ensures that immediate help is dispatched to the scene quickly and efficiently. NPSBN e9-1-1 service is available on NPSBN for both primary and secondary users. The NPSBN emergency service follows 3GPP standards (60) in addition to other industry-established and emerging emergency service practices (61) (62).

Since the advent of the IP-based broadband services, the 9-1-1 services have evolved significantly from 9-1-1 to e9-1-1 (Enhanced 9-1-1) and now NG 9-1-1 (Next Generation 9-1-1). The primary difference between 9-1-1 and Enhanced 9-1-1 (e9-1-1) service is that e9-1-1 provides improved location information for emergency calls. While standard 9-1-1 allows callers to reach emergency services, e9-1-1 automatically transmits the caller's location to dispatchers, which can significantly reduce response times. This system can determine the caller's location through technologies such as GPS or by triangulating signals from cell towers ensuring that help can arrive more quickly and efficiently.

NG9-1-1 is an advanced nationwide initiative based on NG9-1-1 NENA and other standards. NG9-1-1 enables seamless communication between an emergency caller, dispatchers, and first responders. The NG9-1-1 infrastructure incorporates innovative technology and cloud-based IP networks that enable emergency callers in need to send potentially life-saving information through voice and multimedia capabilities. NG9-1-1 provides PSAPs and first responders with 9-1-1 caller location data that is more accurate and in real time compared to e9-1-1 which only shows emergency caller's location. Thus, NG9-1-1 enables first responders to locate citizens in need in emergencies with less guesswork, significantly reducing response time and saving lives.

The additional goal of the NG9-1-1 initiative is for PSAPs and other public safety organizations across the country to upgrade their current 9-1-1 systems to provide communities with faster and more accurate emergency response services.

It is important to note that PSAPs are not within the scope of FNA and the NPSBN. PSAP is a physical location where 9-1-1/e9-1-1 emergency telephone calls are delivered to and answered for a defined geographic service area. In the United States, PSAPS are typically managed and owned by local, state, or regional governments who are responsible for handling emergency calls and routing them to the appropriate public safety agencies. The specific features and capabilities of individual PSAP and its ownership can vary by jurisdiction, with operations often overseen by local law enforcement or emergency management departments. Many of these PSAPs are still using old technology and are not capable of getting full advantages of NPSBN features due to lack of funding and technical knowledge.

9.14.1 NPSBN e9-1-1 Service Architecture

NPSBN e9-1-1 service is a critical emergency feature for first responders. An emergency can happen any time and in any location with anyone which must be fully supported to save lives and properties. Figure 9.18 provides significantly complex emergency response systems functionalities in a simplified architecture. The figure illustrates an emergency caller dialing e9-1-1. The

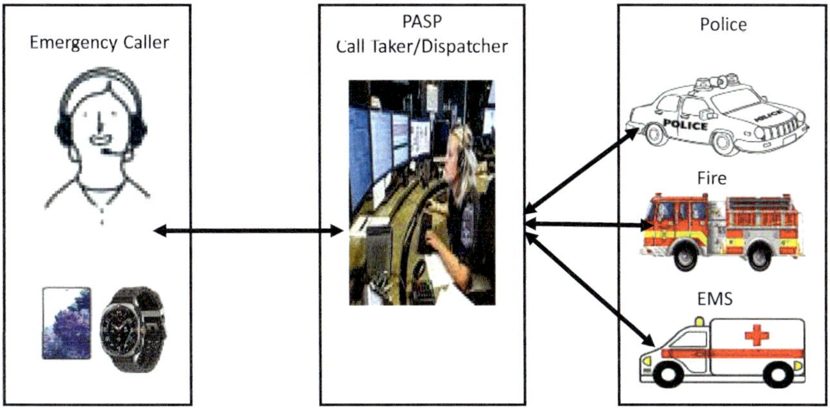

FIGURE 9.18
NPSBN e9-1-1 Simplified Architecture.

call is received by a call taker at PSAP. The call taker gets the appropriate information about the nature of the emergency, locations, etc., and makes the decision to send police, fire, or emergency services. As stated earlier, PSAPs ownership can vary by jurisdiction, size, and technologies. In a small size PSAP, a call taker performs both the task of call taking and dispatching. In a larger PSAP, the call taking and dispatching functions are generally performed by individual groups. In such cases, the call taker hands over the appropriate information to the dispatcher who performs the dispatching function appropriately

9.14.2 NPSBN e9-1-1 Services

Figure 9.19 illustrates e9-1-1 functional architecture describing the end-to-end e9-1-1 call procedure. The NPSBN e9-1-1 supports location information of the emergency caller so that PSAP dispatcher provides the correct location to first responders to provide emergency help promptly and efficiently. Figure 9.19 illustrates the NPSBN e9-1-1 functional architecture.

The figure illustrates end-to-end 9-1-1 call subsystems based on 4G LTE architecture. A caller initiates a 9-1-1 call over his device connecting to NPSBN RAN, EPC, and IMS core. Caller location information is provided byGateway Mobile Location Center (GMLC) and passed to PSAP. PSAP call receiver/dispatcher gets additional information from the caller, assesses the nature of the emergency, and dispatches emergency personnel.

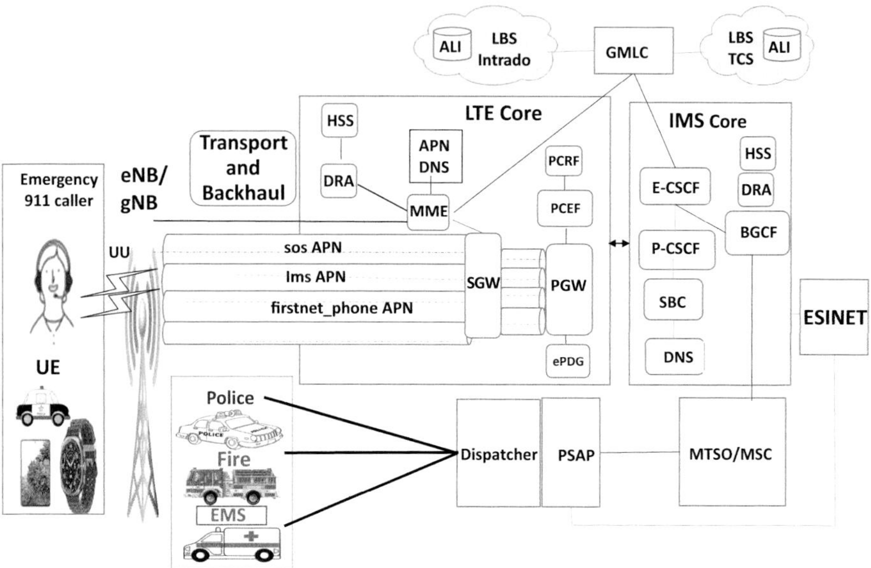

FIGURE 9.19
NPSBN e9-1-1 Functional Architecture.

NPSBN utilizes an SOS APN for e9-1-1 service implementation on NPSBN EPC. NPSBN shares the location database with FN partner, AT&T. Currently, AT&T uses geographic locations based on market service code and has two GMLC providers, Intrado and TCS, which provide location services and manage PSAP connections. PSAP call back is also supported. NPSBN 9-1-1 service design also supports text to 9-1-1 service. FN/AT&T is also actively investigating indoor location accuracy enhancements incorporating Z-axis features.

In the NPSBN IMS, the Location Retrieval Function (LRF) retrieves location information for users that have initiated an emergency session. This location information is provided to the Emergency CSCF (E-CSCF) to facilitate routing of the call to a suitable PSAP.

In location services, Automatic Location Information (ALI) provides the dispatcher with the precise location of a caller who has dialed 9-1-1. This information is automatically retrieved from the caller's telephone number and helps emergency services to quickly determine the location from which the call originated, which is critical for timely response.

Emergency Services IP Network (ESInet) is an IP-based network designed to support NG9-1-1 systems interoperability and transport. ESInets enable the transport of emergency calls and associated data, such as location information, text messages, photos, and videos, to PSAPs.

The following are the high-level interactions among different nodes involved in emergency 9-1-1 calls:

- FN users initiate emergency 9-1-1 call from FN 4G/5G-enabled device (UE).
- eNB/gNB interfaces over S1 C to MME.
- MME delivers the location information along with IMSI/IMEI to GMLC.
- MME interface with SGW. Using sos APN, ims APN, and firstnet_phone APN, secured tunnels are created between eNB/gNB and SGW/PGW.
- FN P-CSCF forwards the call to E-CSCF.
- E-CSCF retrieves Emergency Service Routing Key (ESRK) and forwards calls to BGCF with ESRK. An ESRK is a 10-digit number used in the North American Numbering Plan to uniquely identify a wireless emergency call. The ESRK helps route the call through the network to the appropriate PSAP and is used to retrieve associated ALI data.
- BGCF routes the call to appropriate Mobile telephone Switching Office / Mobile Switching Center (MTSO/MSC) based on the ESRK.
- MTSO/MSC routes to Signaling Reduction (SR) which routes it to the correct PSAP.
- PSAP dispatchers connect to first responders, depending on the emergency situations to police, fire, and EMS.
- First responder has emergency caller information along with the location and the emergency.

9.14.3 NPSBN e9-1-1 Service Design and Call Flow

Emergency bearer services functionalities are provided by the serving network when the network is configured to support emergency services. Emergency bearer services are provided based on the normal or emergency-attached UEs, the status of UEs on the network, i.e., whether their UE is in the limited-service or a full-service state. A limited-service state occurs when a mobile device (UE) successfully attaches to the network but lacks full access to communication services. This can happen if the device is registered but does not have proper access rights, such as being outside the coverage of a full-service area or due to subscription issues. In this state, the device typically can send only emergency calls or use specific services defined by the network operator. Receiving emergency services in a limited-service state does not require a subscription. Different behaviors of emergency bearer support regulated by local regulations are as follows:

a. **Valid UEs only.** No limited-service state UEs are supported in the network. Only UEs that have a valid subscription and are authenticated and authorized for PS service in the attached location are allowed. UEs should be attached to the network and then perform a PDN connection request when an IMS emergency session is detected by the UE.
b. **Only UEs that are authenticated allowed.** These UEs must have a valid IMSI. These UEs are authenticated and may be in limited-service state due to being in a location that they are restricted from service. A UE that cannot be authenticated will be rejected.
c. **IMSI required, authentication optional.** These UEs must have an IMSI. If authentication fails, the UE is granted access and the unauthenticated IMSI retained in the network for recording purposes. The IMEI is used in the network as the UE identifier. IMEI-only UEs will be rejected (e.g., UICCless UEs).
d. **All UEs allowed.** Along with authenticated UEs, this includes UEs with an IMSI that cannot be authenticated and UEs with only an IMEI. If an unauthenticated IMSI is provided by the UE, the unauthenticated IMSI is retained in the network for recording purposes. The IMEI is used in the network to identify the UE.

Emergency Bearer Service

To provide emergency bearer services, the MME is configured with MME emergency configuration data that is applied to all emergency bearer services that are established by an MME on UE request. The MME emergency configuration data contains the emergency APN, which is used to derive a PGW, or it may contain the statically configured PGW for the emergency APN.

Attach to SOS APN (EPS View)

Note that an e9-1-1 call in Wi-Fi is a last-resort choice by the device. Any RAN access, including other wireless carriers, is preferred over Wi-Fi.

In the 9-1-1 emergency response system, ALI provides the dispatcher with the precise location of a caller who has dialed 9-1-1. This information is automatically retrieved from the caller's telephone number and helps emergency services to quickly determine the location from which the call originated, which is critical for timely response.

ERSK and Emergency Service Routing Digit (ESRD) for eCGI locations:

- FN users initiate emergency 9-1-1 call.
- MME delivers the eCGI, along with IMSI/IMEI to GMLC.
- GMLC uses the information to map ALI/SR database for precise location of a caller who has dialed 9-1-1. ALI also incorporates GPS information provided by mobile phones during a 9-1-1 call. When a call is made, the phone's GPS capabilities can help determine the caller's location, which is then transmitted to emergency services to enhance response accuracy.
- FN P-CSCF forwards the information to shared AT&T E-CSCF.
- E-CSCF retrieves ESRK and forwards the call to BGCF with ESRK.
- BGCF routes the call to appropriate MTSO/MSC based on the ESRK.
- MTSO/MSC routes the SR which further routes it to the correct PSAP.

3GPP text-SR and Idle Mode Signaling Reduction (ISR) functions in emergency LTE networks are designed to optimize and reduce signaling traffic, particularly during emergency situations where network resources are strained.

ISR allows devices to maintain their context in a more efficient manner, minimizing the signaling required when transitioning between idle and connected states. SR focuses on reducing the overall signaling load within the network by limiting unnecessary messages and improving responsiveness during high-demand situations such as emergencies.

These functions improve the reliability and efficiency of communication in critical scenarios, ensuring that emergency services can operate effectively.

9.14.4 NPSBN Emergency 9-1-1 Call Flow for Location Information

First responders' crucial responsibilities to safeguard lives and properties during the emergency response require that PSAP dispatchers provide the precise location information of the e-9-1-1 caller to first responders. Figure 9.20 illustrates the high-level call flows and the steps performed by different segments of the NPSBN network to support location information.

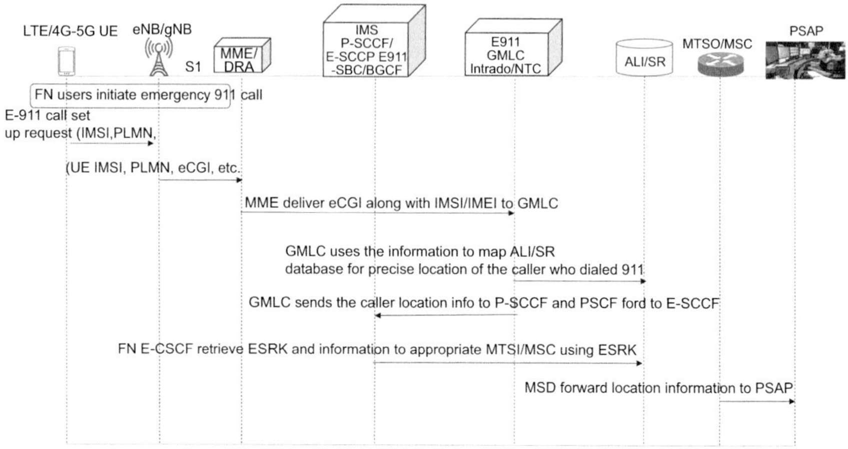

FIGURE 9.20
NPSBN Emergency 9-1-1 Call Flows.

9.14.5 Emergency 9-1-1 Service QPP and Performance

It is crucial that emergency 9-1-1 call be delivered efficiently to PSAP call takers/dispatchers, enabling them to send emergency team to assist the 9-1-1 caller. As stated earlier, emergency signalling must be handled promptly and user information, e.g., location, voice, video text, and data must be promptly made available to first responders.

The PCRF provides the PGW with the QoS parameters, including an ARP value reserved for the emergency bearers to prioritize the bearers when performing admission control. The PCRF ensures that the emergency PGW connection is used only for IMS emergency sessions. The PCRF rejects an IMS session established via the emergency PGW connection if the Administrative Function (AF) (i.e., P-CSCF) does not provide an emergency indication to the PCRF.

The default and dedicated EPS bearers of a PGW connection associated with the emergency APN are reserved for IMS emergency sessions and do not allow any other type of traffic. The PGW shall block any traffic that is not from or to the addresses of network entities (e.g., P-CSCF) providing IMS emergency services. This functionality is used during the attach procedure and during the UE-requested PGW connectivity procedure, in both cases when establishing emergency bearer services.

The UE uses Access Class 10 and Emergency RRC cause code. The emergency service does not require SIM or HSS authorization and the MME has local emergency APN profile.

It should also be noted that generally local regulations require supporting calls from a caller who may not have subscription data in the MME. Additionally, the local network may want to provide IMS emergency services support differently than what is allowed by the UE subscription. Therefore, the initial QoS values used for establishing emergency bearer services are configured in the MME within the MME emergency configuration data.

Table 9.8 provides a high-level view of QoS and other performance-related values assigned for the e9-1-1 calls by the standards bodies.

Detailed discussion of the QPP (e.g., Access Class, QCI, ARP, etc.) are discussed in Chapter 10, "NPSBN Quality of Service and Performance Chapter."

9.14.6 e9-1-1 Service – Key Points

Some of the key features of NPSBN E9-1-1 service are

- E9-1-1 calls made by FN primary users, extended primary users or secondary/individual users are treated equally.
- All the E-9-1-1 location functions like P-LRF and ESMLC are shared between NPSBN and AT&T networks.
- Dedicated NPSBN nodes will be configured to point to appropriate P-LRF and ensure MME/DRA appropriate routing in place.
- Dedicated NPSBN nodes will be configured to ensure interface between MME and SMLC.
- eCGI of shared eNB/gNB is used. eCGI assignment follows the conventional macro eNB/gNB eCGI assignment rules and uses shared market service code.
- Any ongoing e9-1-1 calls on B14 will not be preempted.

TABLE 9.8

NPSBN Emergency 9-1-1 QPP

Emergency Services	Access Class	APN and Bearer Type	QCI	ARP	PCI	PVI
VoLTE – Emergency Signalling	10	sos – default bearer	5	7	Enabled	Disabled
VoLTE – Emergency Voice media	10	sos – dedicated bearer	1	7	Enabled	Enabled

9.15 WIRELESS EMERGENCY ALERTS

A Wireless Emergency Alert (WEA) is an NPSBN feature that enables government authorities to send urgent messages through mobile networks to alert the public about emergencies. This can include severe weather warnings, local emergencies requiring immediate action, America's Missing: Broadcast Emergency Response (AMBER) alerts for missing children, and presidential alerts during a national emergency. WEAs are designed to be short, concise, and direct to ensure that critical information reaches as many people as possible. WEA specifications and the processes are provided in Part 10 of the Code of Federal Regulations Title 47 (63) (www.ecfr.gov/current/title-47/chapter-I/subchapter-A/part-10). Although, the WEA is currently on volunteer basis and mobile service providers may elect to transmit alert messages to the public, NPSBN has implemented the WEA service for all FN users.

9.15.1 WEA Service Architecture

WEA architecture describes the system and protocols used to disseminate emergency messages to the public through wireless communication technologies. It is designed to send time-sensitive information about emergencies broadcasted by federal, state, and local agencies. The WEA architecture involves a partnership between various organizations, including the Federal Emergency Management Agency (FEMA), mobile carriers, e.g., NPSBN, and public safety officials, to ensure alerts can effectively reach consumers via their mobile devices. The WEA high-level logical architecture is illustrated in Figure 9.21.

Each government agency sends the specific emergency alert to the FEMA Alert Aggregation System (AAS) where the alerts are combined and sent to a FEMA's Integrated Public Alert and Warning System Gateway (IPAWS GW) for onward transmission to NPSBN gateway. Finally, NPSBN receives the alert and broadcasts to all NPSBN users on their UE.

9.15.2 WEA Service Design and Call Flow

FEMA's IPAWS GW provides the infrastructure for the dissemination of alerts and warnings to the public through various communication channels, such as radio, television, and mobile devices. It enables authorized agencies

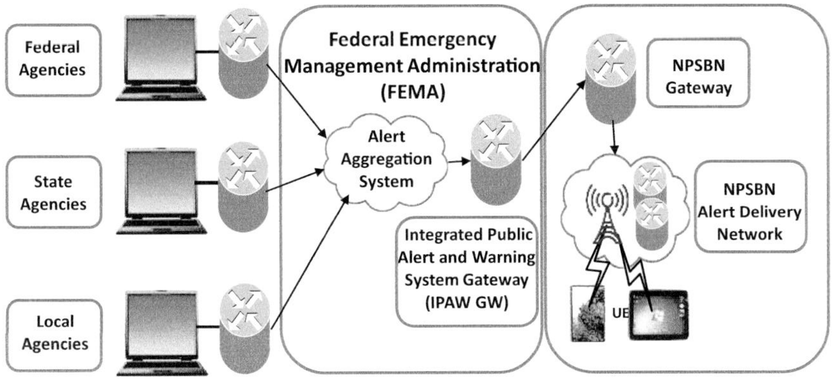

FIGURE 9.21
WEA Logical Architecture.

to issue timely and effective alerts about emergencies, including severe weather events, natural disasters, and other public safety threats. FEMA uses the WEA's Common Alert Protocol (CAP) to aggregate and send the alert to NPSBN for onward broadcast to NPSBN users.

FEMA's IPAWS Design (64) details are illustrated in Figure 9.22. FEMA uses CAP, which is also used by all other agencies and the commercial mobile networks including NPSBN for the WEA. Whether to broadcast it nationally or regionally is dependent on the geographic area of the alert specified by the GW in its transmission of the alert to the AT&T's Broadcast Messaging Center (BMC) network node to transmit to the FN user devices. FEMA manages and funds the core IPAWS infrastructure but not the alert origination tools or distribution channels. IPAWS is designed to provide a single authenticated entry point to the Emergency Alert System (EAS), WEA, National Oceanic and Atmospheric Administration (NOAA), Weather Radio All Hazards System, and two other alerting systems. Some communities use older emergency notification systems, which may require an interface to IPAWS for interoperability.

9.15.2.1 Alert Delivery Systems

Table 9.9 describes the various categories of emergency alert and their delivery systems either by "Pull" or "Push" system.

The push delivery system allows authorized users to send alerts or warnings directly to the IPAWS platform, which then disseminates the information to

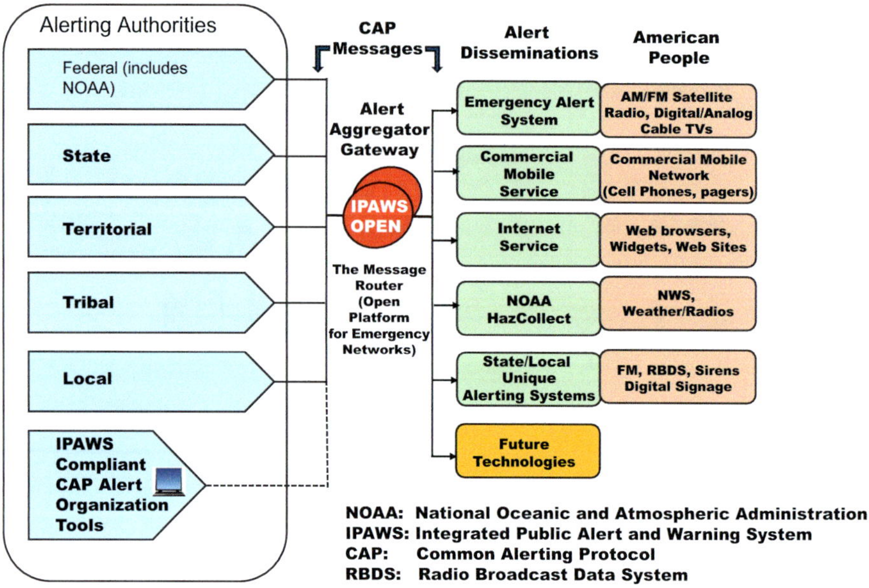

FIGURE 9.22
WEA Component Connectivity Design.

the public via various channels, such as radio, TV, and cell towers. In contrast, the pull delivery system enables users to access and retrieve alerts from the IPAWS system on demand, ensuring they can obtain timely information as needed.

9.15.3 WEA Service QPP and Performance

FEMA's WEA includes specific performance requirements to ensure effective delivery of emergency information. These requirements address aspects like alert message length, transmission time, geographic targeting, and user opt-in, aiming for rapid dissemination of critical alerts such as severe weather warnings, local emergencies, and AMBER alerts. Additionally, the Quality Performance Planning requirements involve benchmarks for alert message success rates, system resilience, and stakeholder engagement. The following are the specific WEA requirements.

NPSBN/AT&T is required to participate in Required Monthly Test (RMT). RMT is initiated by the FEMA via IPAWS GW using a predefined test

TABLE 9.9
WEA Delivery Systems

Alert Type	Delivery System	Notes
Emergency Alert System	Pull	Not required to broadcast local message
Weather Emergency Alert	Push	Limited message length
National Weather Service Radio	Push	Additional permissions required
All Hazards Information Feed	Pull	Varies by systems

message at the specific day/time of each month. A participating CMS provider (NPSBN/AT&T) must distribute an RMT in its WEA coverage area within 24 hours of reception.

State and local tests are random in nature and can be initiated by authorized state and local agencies. The CMS provider must transmit these test alerts with no delays the same way as any real alerts are transmitted.

9.15.4 WEA – Key Points

The FEMA's WEA is an important aspect of the NPSBN communication to FN users. The WEA key features are summarized here:

- The WEA is a volunteer service provided by NPSBN to its users.
- The WEA can be triggered by any federal, state, local, tribal, and territorial government agencies relating severe weather warnings, local emergencies requiring immediate action, AMBER alerts for missing children, and presidential alerts during a national emergency, etc., via FEMA's IPAWS GW using CAP.
- FEMA manages and funds the core IPAWS infrastructure but not the alert origination tools or distribution channels. IPAWS is designed to provide a single authenticated entry point to the EAS.
- NPSBN complies with FEMA's WEA specifications and supports the WEA. Alerts are typically disseminated within minutes of an incident occurring. However, the actual delivery time can vary based on several factors, including the type of alert, the location, and the network conditions. Generally, alerts are intended to reach mobile devices within seconds to a few minutes after being issued.

9.16 WIRELESS PRIORITY SERVICES

Wireless Priority Service (WPS) is a federal program that authorizes to prioritize calls over wireless networks and provide priority access to certain government organizations including federal, state, local, and tribal police department, fire department, public safety answering point or 9-1-1 call center, EMS, essential healthcare provider, or any other organization that uses telecommunication services necessary for the public health, safety, and maintenance of law and order particularly during emergencies or disaster situations. NPSBN facilitates WPS prioritization services to all users who are approved to receive WPS service.

The U.S. Department of Homeland Security (DHS) is the sole authority to approve the WPS enrollment application of the individual submitting and certifying that his/her work responsibilities meet the mission of national security and emergency preparedness requirements as set forth in the WPS program information. If the organization qualifies, the WPS program office will contact NPSBN/AT&T and complete the process. There is no charge to WPS subscribers for enrollment or use.

The method to invoke WPS on the WPS-enabled device is to dial *272 and then enter the phone number the user wants to call. This will initiate a priority calling queue. WPS calls do not preempt emergency calls in progress but provide priority status over cellular communications networks. The WPS program is in effect all the time and is not contingent on a major disaster or attack to take place to get WPS service.

9.16.1 FCC WPS Priority Criteria

The FCC has specified priority levels and qualifying criteria (65) for WPS as given here:

Priority 1: Executive Leadership and Policy Makers users who qualify for the Executive Leadership and Policy Makers priority are assigned priority one. Examples of those eligible users include (i) the President of the United States, the Secretary of Defense, select military leaders, and the minimum number of senior staff necessary to support these officials; (ii) State governors, lieutenant governors, cabinet-level officials responsible for public safety and health, and the minimum number of senior staff necessary to support these officials; and (iii) Mayors, county commissioners, and the minimum number of senior staff to support these officials.

Priority 2: Disaster Response/Military Command and Control users who qualify for the Disaster Response/Military Command and Control. Individuals eligible for this priority include the personnel who manage the initial response to an emergency at the local, state, regional, and federal levels. The personnel selected for this priority should be responsible for ensuring the viability or reconstruction of the basic infrastructure in an emergency area. In addition, personnel essential for the continuity of government and national security functions (such as the conduct of international affairs and intelligence activities) are also included in this priority. Examples of those eligible include (i) Federal emergency operations center coordinators, e.g., Manager, National Coordinating Center for Telecommunications, National Interagency Fire Center, Federal Coordinating Officer, Federal Emergency Communications Coordinator, Director of Military Support; (ii) State Emergency Services Director, National Guard Leadership, State and Federal Damage Assessment Team Leaders; (iii) Federal, state and local personnel with continued government responsibilities; (iv) Incident Command Center Managers, local emergency managers, other state and local elected public safety officials; and (v) Federal personnel with intelligence and diplomatic responsibilities.

Priority 3: Public Health, Safety, and Law Enforcement Command users who qualify for the Public Health, Safety, and Law Enforcement Command priority will be assigned priority three. Eligible for this priority are individuals who direct operations critical to life, property, and maintenance of law and order immediately following an event. Examples of those eligible include (i) Federal law enforcement command; (ii) State police leadership; (iii) Local fire and law enforcement command; (iv) Emergency medical service leaders; (v) Search and rescue team leaders; and (vi) Emergency communications coordinators.

Priority 4: Public Services/Utilities and Public Welfare users who qualify for the Public Services/Utilities and Public Welfare priority will be assigned priority four. Eligible for this priority are those users whose responsibilities include managing FCC 00-242 public works, utility infrastructure damage assessment, restoration efforts, and transportation to accomplish emergency response activities. Examples of those eligible include (i) Army Corps of Engineers leadership; (ii) Power, water, and sewage and telecommunications utilities; and (iii) Transportation leadership.

Priority 5: Disaster Recovery Users who qualify for the Disaster Recovery priority will be assigned priority five. Eligible for this priority are those individuals responsible for managing a variety of recovery operations after the initial response has been accomplished. These functions may include managing medical resources such as supplies, personnel, or patients in medical facilities. Other activities such as coordination to establish and stock shelters, to obtain detailed damage assessments, or to support key disaster field office personnel may be included. Examples of those eligible include (i) Medical recovery operations leadership; (ii) Detailed damage assessment leadership; (iii) Disaster shelter coordination and management; and (iv) Critical Disaster Field Office support personnel.

9.16.2 WPS Architecture

The WPS utilizes the NPSBN VoLTE service architecture, except WPS-specific QPP are added to comply with *272 dialing priority instead of general VoLTE QPP.

- FCC Order 00-242 WPS priority 1 and 2 users have higher priority than FirstNet primary users. Other WPS users have the same Access Class, ARP Priority Level and Preemption capabilities.
- FCC Order 00-242 addresses the priority of wireline service providers (WSPs) during periods of increased demand, such as during emergencies. The order aims to ensure the reliability of communications by prioritizing certain types of communications and signaling mechanisms, allowing WSPs to manage their resources effectively. It establishes guidelines for service providers to enhance communication services to ensure *272 (WPS) reliability.

9.16.3 WPS Design and Call Flow

Like WPS architecture, the WPS design utilizes VoLTE design, utilizes the NPSBN VoLTE call flows, and WPS FCC Order 00-242-specified WPS priorities. NPSBN supports WPS features at all segments including the handset (UE), Radio Access, Core, and IMS. IMS APN (voice) is used for the WPS. Additionally, when the user makes a WPS, by appending *272, the end-to-end priority treatment is invoked, not only in the EPC/RAN, but across the IMS network and extended to the VoIP and TDM networks. VoLTE-to-VoLTE calls will terminate within IMS/EPC/RAN or across Network-to-Network Interface (NNI) with WPS capable carriers.

9.16.4 WPS QPP and Performance

NPSBN ensures that the WPS users get the WPS services priority QPP specified by FCC Order 00-242. For WPS, QCI 1 and ARP 1 values are used for conversational voice (high priority).

9.16.5 WPS – Key Points

- NPSBN supports FCC WPS FCC Order 00-242 priority assignment criteria in the entire 4G/5G network segments.
- WPS QoS parameters that FN WPS service has assigned currently cannot be guaranteed in the roaming partner's network.
- For the WPS, QCI 1 and ARP 1 values are used for conversational voice.

9.17 COMMUNICATIONS ASSISTANCE FOR LAW ENFORCEMENT ACT

The Communications Assistance for Law Enforcement Act (CALEA) is a statute enacted by Congress in 1994 to require that telecommunications carriers and manufacturers of telecommunications equipment design their equipment, facilities, and services to ensure that they have the necessary surveillance capabilities to comply with legal requests for information.

NPSBN supports CALEA and provides access to the data requested by the law enforcement authorities as per CALEA requirements to meet the following objectives:

- CALEA requires a "telecommunications carrier," as defined by the CALEA statute, to ensure that equipment, facilities, or services that allow a customer or subscriber to "originate, terminate, or direct communications," enable law enforcement officials to conduct electronic surveillance pursuant to court order or other lawful authorization.
- CALEA is intended to preserve the ability of Law Enforcement Agencies (LEA) to conduct electronic surveillance by requiring that telecommunications carriers and manufacturers of telecommunications equipment design and modify their equipment, facilities, and services to ensure that they have the necessary surveillance capabilities as communications network technologies evolve.
- CALEA network converges to national gateways based on service type for delivery to law enforcement.
  - Voice Core (VoLTE, Video Calling)
  - Lawful Intercept-Packet Data (LI-PD)
  - Internet Access and Services – Messaging, converged services, lawful location, etc.
  - Surveillance provisioning for NPSBN EPC will be performed via Electronic Law Enforcement Activity Tracker (ELEAT).

9.17.1 CALEA Service Architecture

CALEA requires (66) NPSBN to upload information requested by LEA to a CALEA database enabling the LEA to receive the requested information for their review and legal action. Additionally, LEA also performs active surveillance of an individual or group of people duly approved by the U.S. court. Figure 9.23 provides a simple CALEA logical architecture.

As illustrated in the figure, CALEA requests NPSBN to provide information relating to VoLTE, video calls, data packets, messaging, and internet data sent to the monitoring database for LTE review and action.

9.17.2 CALEA Service Design and Call Flow

CALEA compliance (67) requires NPSBN to deliver the specific details of emails, internet access services, packet data, messaging, and other related

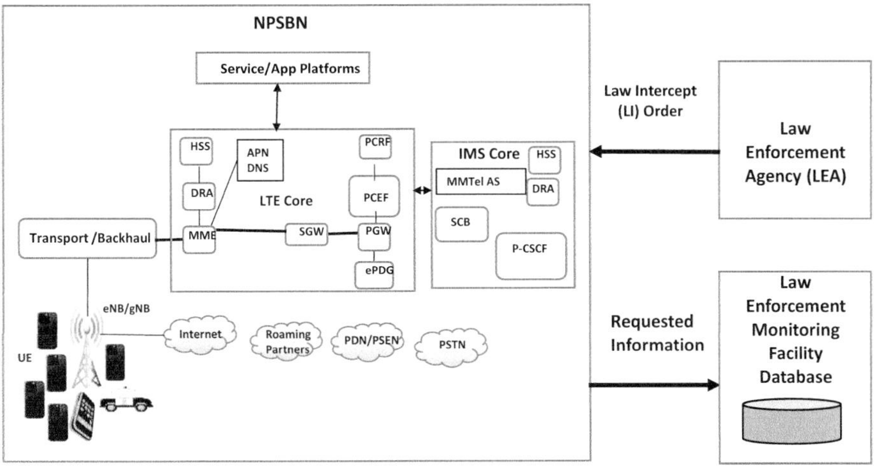

FIGURE 9.23
NPSBN CALEA Logical Architecture.

information including specifications for IP delivery (68). Figure 9.24 illustrates the CALEA service component design to facilitate CALEA compliance.

As illustrated in the figure, Mediation Functions (MD) in CALEA compliance refer to the processes and systems that ensure LEA can access lawful communications and data in a manner that respects privacy laws while adhering to legal requirements. These functions facilitate the interception of telecommunications and data while managing the complexities of maintaining user confidentiality and ensuring that the interception is performed only in compliance with legal mandates. NPSBN mediation system enables software to capture and reroute the traffic of a named suspect. The Delivery Function (DF) is a standards-based interface to deliver the required information to be collected into LEA repository. AF includes the establishment of policies, procedures, and practices that promote effective management, support operational efficiency, and maximize public safety while ensuring accountability and transparency. Administratively, it involves maintaining documentation, conducting self-assessments, and participating in accreditation reviews.

NPSBN works with LEA to establish secure interfaces to exchange information to LEA via HI interface. The HI1, HI2, and HI3 are specific interfaces between NPSBN and LEA. It is designed to provide law enforcement with access to the CDR and other non-content information necessary for investigations, ensuring compliance with legal standards for surveillance. The X1, X2, and X3 interfaces are standard interfaces used by law enforcement with wiretap capabilities for voice and data communications. It defines

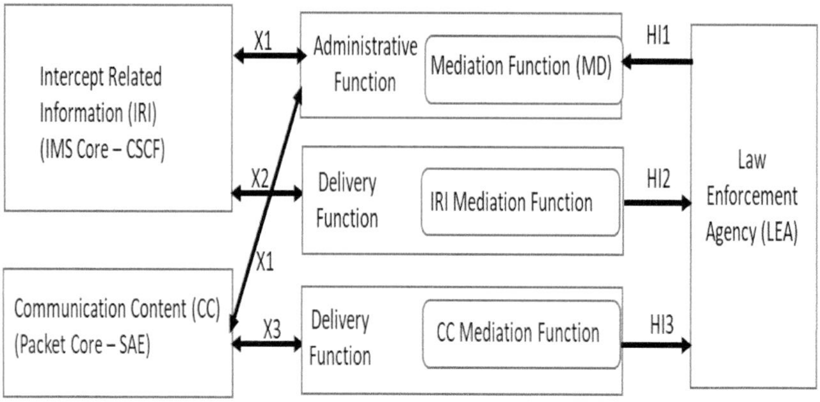

FIGURE 9.24
NPSBN CALEA Service Component Design.

how law enforcement can access communications data during investigations while ensuring that privacy and security principles are maintained.

9.17.3 CALEA Service QPP and Performance

There are no specific QoS requirements specified for delivery of the content required by CALEA. QoS, integrity of the delivered IRI, and CC information are based on agreement between the relevant authorities and NPSBN/AT&T. However, the NPSBN-specific QPP will be applicable, as necessary. The IRI and CC will be delivered in as near real time as possible.

9.17.4 CALEA Service – Key Points

The following are the key features of CALEA service:

- The intent of the NPSBN CALEA service is to comply with the LEA and support necessary surveillance capabilities and provide requested information.
- CALEA is not intended as a service for the NPSBN regular user. Therefore, it is not a user-specific service for their day-to-day work.
- CALEA's "safe harbor" is a designated secure location where law enforcement can conduct surveillance on electronic communications without being obstructed or challenged by the service provider or other entities.

This area ensures that law enforcement can monitor communications as per legal requirements while maintaining the integrity and confidentiality of the investigation.
- Surveillance provisioning for NPSBN EPC will be performed via ELEAT.
- CALEA compliance includes voice services (VoLTE, Video call), messaging, emails, broadband (packet data), and other information required by the LEA which have obtained legal approval to acquire the information.
- The IRI and CC will be delivered in as near real time as possible.

9.18 3GPP LTE V2X (Vehicle to Everything) Services

3GPP LTE V2X (vehicle to everything)-Definitions
Vehicle-to-Vehicle (V2V) service, often referred to as Cellular Vehicle-to-Everything (C-V2X) or simply V2X, is a technology that allows vehicles to communicate with each other and with other entities like infrastructure, pedestrians, and the 4G/5G networks. 3GPP is developing V2X requirements covering:

- V2V: covering LTE-based communication between cellular vehicles.
- V2P: covering LTE-based communication between a vehicle and a device carried by an individual (e.g., handheld terminal carried by a pedestrian, cyclist, driver, or passenger).
- V2I: covering LTE-based communication between a vehicle and a road-side unit.

3GPP has defined the user cases and intends to continue the architecture and development specifications into 3GPP 5G programs. Also, U.S. Government (DOT and NHSTA-led) are also working on V2X project over 802.11p. Hopefully, both efforts would be aligned to develop a single and comprehensive V2X architecture requirements and development activities see Table 9.10.

9.19 SUMMARY OF NPSBN SERVICES

TABLE 9.10

NPSBN List of S ervices

NPSBN Services	Current Status on 4G/5G Networks
Voice Service (VoLTE, Video Call, Voice Mail)	Y
IPME	Y
IR94 Video Calling	Y
Real Time Text	Y
Emergency Voice 9-1-1	Y
Outbound Roaming	Y
Number Sync	Y
Visual Voice Mail	Y
Mission Critical Services (MC Voice, MC Data, MC Video)	Y
Mobile Broadcast Multicast Service (MBMS)	Y
Commission on Accreditation for Law Enforcement Agencies (CALEA)	Y
Wireless Emergency Alerts (WEA)	Y
Wireless Priority Services (WPS)	Y
FirstNet QPP	Y
Internet of Things (IoT)	Y

10

NPSBN Quality of Service and Performance

NPSBN Quality of Service (QoS) and network performance are related concepts but not the same. QoS refers to the overall fulfilment of National Public Safety Broadband Network (NPSBN) service as perceived by the users. QoS encompasses a broader set of factors that influence the user's experience. NPSBN performance, on the contrary, typically describes how well NPSBN operates in terms of availability, reliability, mean time to repair, etc. Both are crucial for assessing user satisfaction and service effectiveness.

Quality and performance in public safety networks are crucial as they directly impact the effectiveness and reliability of emergency response services. High-quality communication networks ensure that first responders share critical information quickly and without interruption, which can save lives during emergencies. Performance measures such as availability, reliability, and capacity are essential to handle the demands of public safety operations, especially during large-scale incidents or disasters. Furthermore, quality assurance helps maintain the integrity and security of sensitive data transmitted across these networks.

10.1.1 NPSB QOS FOUNDATION

NPSBN supports significantly rigid QoS criteria, enabling first responders to perform their life-saving duties efficiently and without any undue interruptions (69, 70). The criteria are as follows:

DOI: 10.1201/9781003557739-12

10.1.1.1 Access Priority Criteria

- Access Class
  Access Class (AC) is a feature that categorizes mobile devices into different groups based on their priority for accessing the network. Each device is assigned to one of the 10 ACs (ranging from 0 to 9), where lower numbers typically denote higher priority. AC is assigned to each user to indicate its access priority and stored in the SIM. This classification helps manage network access, especially during high traffic situations, by determining which devices can connect to NPSBN.
- Access Class Barring
  Access Class Barring (ACB) is a mechanism used to restrict access to NPSBN during peak usage times. It allows the network to limit the number of devices from certain access classes that can access the network, effectively preventing network overload. ACB applies to specific scenarios, such as during a call setup or when there is a significant surge in demand. ACB is an eNB/gNB-activated action that reduces the number of connection requests presented to the eNB/gNB.
- High-Priority Access
  High-Priority Access (HPA) in NPSBN refers to the mechanism that allows certain users, typically those classified as having high priority (such as emergency services or special event personnel), to gain quicker access to network resources. This ensures that during high traffic conditions, these users have access to NPSBN resources to perform their public safety duties uninterrupted.
- Overload Control
  In NPSBN overload control manages and mitigates the congestion experienced in the network, ensuring QoS for high-priority users. When the traffic load exceeds the network capacity, generally when the RAN reaches near 80% capacity, overload control strategies such as admission control, traffic shaping, and prioritization of certain types of traffic are employed to maintain service reliability and performance.

  This involves dropping lower-priority traffic or delaying nonessential services to allocate resources effectively for critical communications to support first responders.
- Preemption:
  Preemption in NPSBN is the ability by high-priority users to terminate or "bump" a low-priority user's "session" who has an ongoing service communication in NPSBN. It is essential to maintain QoS for high-priority

users. Also, the process allows NPSBN to dynamically allocate resources based on real-time demand and the priority levels of various users, ensuring that critical communications remain uninterrupted.

10.1.1.2 NPSBN Service Priority Criteria

Service priority criteria are the rules and mechanisms used to prioritize certain types of traffic or services over others and include the following:

10.1.1.2.1 QoS Class Identifier (QCI)

QCI is a key component in NPSBN that defines the level of QoS provided for different types of data traffic. It categorizes the data flow based on the service requirements, such as latency, jitter, packet loss, and bandwidth, enabling the network to manage resources effectively to ensure that applications receive the appropriate level of performance. Each QCI value corresponds to specific QoS parameters that help prioritize traffic, improve user experience, and maintain service reliability. QCI is the attribute of an Evolved Packet System (EPS) bearer indicating service priority on packet level or packet priority. QoS comprises the following elements: (1) Resource (bearer) Type, (2) Priority, (3) Packet Delay Budget, and (4) Packet Loss Rate.

1. NPSBN Bearers

   NPSB supports the following types of bears.

   a. Guaranteed Bit Rate (GBR) Bearer: Reserves guaranteed bandwidth in the network. No expected packet loss during congestion.
   b. Non-Guaranteed Bit Rate (Non-GBR) Bearer: No specific bandwidth allocation – best effort services; packet loss during network congestion
   c. Default Bearer: When user equipment (UE) attaches for the first time, it is assigned a default bearer. The default bearer is the best effort service. QCI 5–9 (non-GBR) are generally assigned to default bearer.
   d. Dedicated Bearer: Provides dedicated tunnel to one or more specific traffic (i.e., VoLTE, video, etc.).
   e. Maximum Bit Rate (MBR) is used for a GBR-type bearer and indicates the maximum bit rate allowed in the LTE network.

f. Aggregate Maximum Bit Rate (AMBR) specifies the maximum upload and download speed associated with the bearer. There are two types of AMBR designation – UE AMBR and Bearer AMBR. UE AMBR is the maximum data rate allocated to a specific UE for its total data transfer, combining all active bearers. Bearer AMBR refers to the maximum data rate that can be allocated to an individual bearer within a user session.

2. Allocation Retention and Priority (ARP)

ARP in NPSBN is used to prioritize and manage the allocation of radio resources to different types of traffic/services based on the desired QoS. ARP consists of two key components: allocation priority and retention priority. Allocation priority dictates how resources are assigned to new requests, whereas retention priority helps maintain service continuity for ongoing sessions under resource constraints. NPSBN conforms to the standards of ARP level: 1–15, 1 being the highest priority level.

Retention priorities are of two types: Preemption Capability Indicator (PCI) and Preemption Vulnerability Indicator (PVI). PCI allows an NPSBN use bearer to "bump" or disconnect other users in case of network congestion, whereas the PVI values identify the specific user vulnerability to be removed from NPSBN.

Table 10.1 provides ARP, PCI, and PVI values and the intended outcome.

3. Packet Delay Budget (including Core Network Packet Delay Budget)

Packet delay or latency budget in NPSBN is the maximum allowable time that a packet can take to travel from Packet Gateway (PGW) to its destination UE. This budget is critical for ensuring that applications requiring low latency, like VoLTE or video streaming, perform satisfactorily. Packet latency accounts for delays from various sources in NPSBN, including processing, transmission, queuing, and propagation delays. It is crucial in emergency communication systems,

TABLE 10.1

Allocation Priority, PCI, and PVI-Intended Outcome on Bearers

Allotment Priority Level	1–9 Depending upon the Service, 1 Being the Highest Priority	
Preemption Capability Indicator (PCI)	1 (Enabled), capable to "bump" another bearer from NPSBN	0 (Disabled), not capable of "bumping" another bearer
Preemption Vulnerability Indicator	1 (Enabled), can be "bumped" by another bearer from NPSBN	0 (Disabled), cannot be "bumped" by another bearer

where real-time information is essential for efficient response and coordination. High latency can negatively impact the effectiveness of communications during critical situations. As per Third Generation Partnership Project (3GPP) specification the packet delay budget is typically 20 S.

4. Packet Loss and Error Rate
   Packet loss or error rate in NPSBN is defined to the percentage of packets that are lost and have failed to reach their destination or received with errors during transmission. This can occur due to network congestion, hardware failures, or poor connectivity. During emergency communications, packet loss can impair the effectiveness of communications, leading to delays or incomplete information, which is particularly detrimental to public safety operations.
5. Jitter
   Jitter measures the variation in packet arrival time, which affects the QoS especially in VoLTE and the consistency of data transmission.
6. Throughput – Download (DL) and Upload (UL)
   Throughput in NPSBN is the amount of data that can be transmitted over NPSBN in a specific period. It is a measure of the network's capacity to handle the volume of communications and data transfers, such as voice calls, text messages, or video feeds that public safety personnel use for efficient operation and coordination during emergencies.
7. Maximum Data Burst Volume (for delay-critical GBR resource type only).
   The maximum amount of data that can be transmitted in a single burst during a data session. This parameter is part of the QoS settings and helps manage network congestion and ensure efficient data transmission.

10.1.2 NPSBN PERFORMANCE FOUNDATION

NPSBN performance indicates how well the NPSBN network segments, service, or system executes its intended functions or tasks. Following are the critical features impacting the NPSBN performance:

1. Availability
   Availability of NPSBN refers to the degree to which the network remains operational and accessible to users, especially in emergency

situations. It implies that the NPSBN can sustain continuous communication and provides reliable service to first responders and emergency personnel, ensuring they can respond effectively when needed. Availability of NPSBN is crucial for first responders to ensure seamless operations during critical incidents. The NPSBN service availability target is 99.99%. The service availability metric will be measured via test probes deployed across the footprint, providing a statistical basis for measurement and regular reporting.

2. Reliability
   The reliability of NPSBN includes factors such as infrastructure, technology, and operational practices. NPSBN aims to achieve utmost reliability to ensure effective communication during emergencies. NPSBN incorporates locally redundant infrastructure and georedundancy to maintain services. NPSBN utilizes highly robust technologies and constantly monitors systems to maintain functionality under stress. However, specific reliability metrics can depend on local implementations and can vary widely. NPSBN reliability conforms to 3GPP specifications recommendation of 99.9999.
3. Retainability
   Retainability in NPSBN is the ability of the network to keep a user connected during mobility, such as when moving between different coverage areas. It indicates the success rate of maintaining an active connection after a handover from one cell to another, ensuring continuous service without drops or interruptions.
4. Mean Time to Repair
   Mean Time to Repair (MTR) in NPSBN is the average time taken to repair a system or component after a failure occurs. It is a critical metric used to assess the reliability and efficiency of a network, indicating how quickly services can be restored to ensure public safety. A lower MTTR signifies a more efficient response to failures, which is crucial in emergency situations.

10.1.3 NPSBN QPP ARCHITECTURE

In NPSBN, Access Class, HPA, QoS, and ARP combined refers to QPP – that is, QoS, Priority, and Preemptions that play vital roles in supporting various NPSBN services and applications. NPSBN QoS architecture is a framework to provide different levels of data transmission quality within 4G/5G networks. It is designed to ensure that various types of service requirements (such as voice, video, data, and emergency calls) are met efficiently. The architecture involves components such as the NPSBN core, which manages QoS parameters, and standardized QoS classes that define the specific requirements for each type of service – including factors such as priority, latency, bandwidth, and error rates. This ensures that critical services receive priority over less critical data transmissions. Also, NPSBN performance framework ensures that the first responders have NPSBN service available, reliable, and ready to access 24 × 7 of 365 days a year.

QPP plays vital roles in supporting various NPSBN services and applications. NPSBN Radio Admission Control (RAC) is a crucial component that manages the allocation of radio resources to UE based on network capacity, QoS requirements, and network congestion. As illustrated in Figure 10.1, based on the UE context, the UE may be "admitted" or "rejected" depending on its Access Class. If the RAN is getting congested, only UE having HPA will be admitted and the bearer will be allocated.

In addition, RAN controls the UE's attachment status based on the PCI and PVI. During the load condition, PCI and PVI play crucial role in continuing the UI attachment or detachment of the UE from the RAN.

QCI, GRB/Non-GRB, default/dedicated bearer assignments are done at the PCRF/PGW based on the NPSBN user profile and appropriate bearer allocation is done at the RNC.

10.1.4 NPSBN QPP DESIGN

NPSBN design incorporates QPP values and their designated characteristics on the NPSBN services and applications.

10.1.4.1 Access Class (AC)

- AC is an identifier assigned to each user to indicate its access priority and is stored in the SIM
- AC 0–9 is assigned for commercial users
- AC 10 is assigned for emergency calls, e.g., 9-1-1 calls
- AC 12 and 13 are assigned for FirstNet primary and extended users

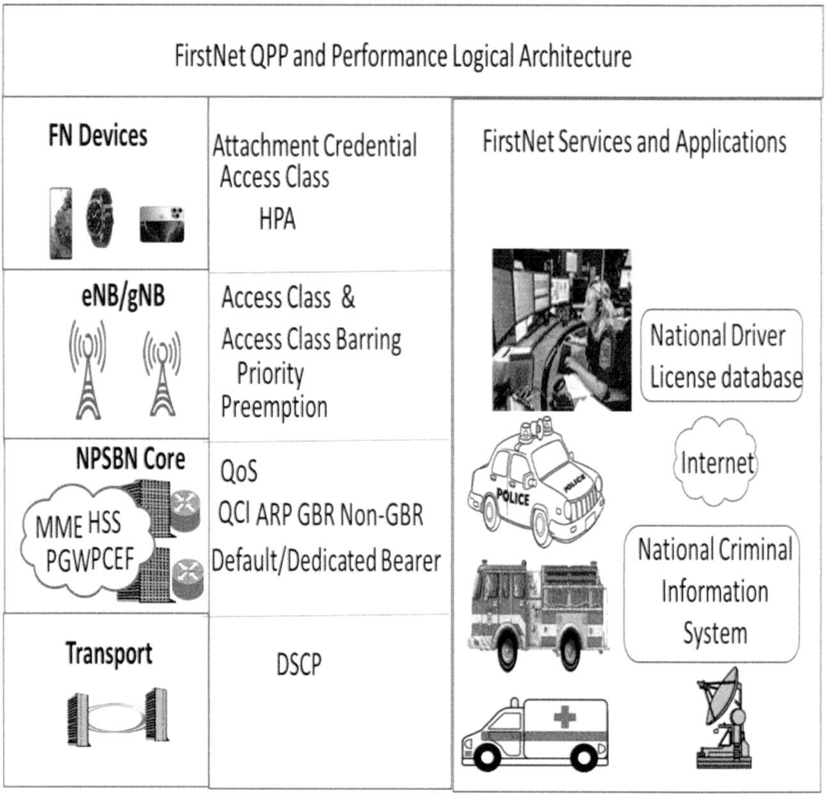

FIGURE 10.1
NPSBN QoS and Performance Logical Architecture.

- AC 14 is assigned for Wireless Priority Services (WPS)
- AC 11 and 15 are for Public Land Mobile Network (PLMN) staff, for operator use

Radio Resource Control (RRC) was established to set HPA if the UE uses AC that is 11–15.

ACB is a RAN-activated control action that reduces the number of connection requests presented to the eNB/gNB.

10.1.4.2 Bearer-Level QoS Parameters

The EPS bearer QoS profile (71) includes the parameters QCI, ARP, GBR, and MBR (72). QoS parameters also include an aggregated set of EPS bearers:

Access Point Name-Aggregated Maximum Bit Rate (APN-AMBR) and UE-AMBR. Each EPS bearer (GBR and Non-GBR) is associated with the following bearer-level QoS parameters:

- QoS QCI
- ARP

In addition, each GBR bearer is associated with:

- GBR
- MBR

Each APN access by a UE is associated with the APN-AMBR QoS parameter. The subscribed APN-AMBR is a subscription parameter stored per APN in the Home Subscribers Service (HSS), which applies as APN-AMBR unless it is modified by the Mobility Management Entity (MME) (e.g., in roaming scenarios) or the PGW, based on local policy or PCRF interactions. The APN-AMBR limits the aggregate bit rate that can be expected across all non-GBR bearers and across all PGW connections of the same APN (e.g., excess traffic may get discarded by a rate-shaping function). Each of those non-GBR bearers could potentially utilize the entire APN-AMBR, e.g., when the other non-GBR bearers do not carry any traffic. The PGW enforces the APN-AMBR in the downlink. Enforcement of APN-AMBR in the uplink is done in the UE and in the PGW.

Table 10.2 provides QCI values for various FN services and applications.

TABLE 10.2

NPSBN QCI Values

Class of Service Indicator (QCI)	NPSBN Service	GBR/ Non-GBR
QCI 1	VoLTE	GBR
QCI 2	Video Call	GBR
QCI 3	Real-Time Interactive Applications	GBR
QCI 4	Stream Video	GBR
QCI 6	Critical Data and Noninteractive Video	Non-GBR
QCI 7	Interactive Video and Voice	Non-GBR
QCI 65	Mission Critical Push to Talk (MCPTT)	Non-GBR
QCI 66	Non-Mission Critical Push to Talk (MCPTT)	Non-GBR
QCI 70	MC Critical Data	GBR

10.1.4.3 NPSBN QoS and DSCP Assignments Mapping

Differentiated Services Code Point (DSCP) is used in NPSBN transport to classify and manage network traffic by defining the level of service quality that different types of data packets should receive (73). DSCP prioritizes packets for various applications, ensuring that time-sensitive services like voice or video maintain better performance across the networks. NPSBN QCI and transport QoS combined are significantly important to meet the desired level of performance set for NPSBN users. Table 10.3 presents the mapping of NPSBN QCI and DSCP. Correct combinations of QCI and DSCP should be assigned at the design time for various NPSBN services and applications to ensure the desire QoS.

10.1.4.4 Summary of NPSBN Service, QPP, and DSCP

NPSBN QPP and DSCP play crucial roles in ensuring effective communication and performance within public safety networks. NPSBN provides dedicated broadband services for first responders, ensuring they have reliable and secure communication. QPP enables prioritization of critical communications ensuring that vital data is transmitted in a timely manner. DSCP is a mechanism used in IP networks to differentiate service classes by marking packets, which ensures in managing bandwidth and reducing latency for high-priority traffic.

Table 10.4 provides a comprehensive summary of the QPP and DSCP including latency and packet loss associated with each NPSBN services and applications. This table provides QPP and DSCP assignment for FN users for

TABLE 10.3

NPSBN QoS and DSCP Mapping

	FirstNet Traffic	
Application Type	**QCI – DSCP**	**MPLS Priority**
VoLTE	QCI 1 (DSCP 46)	Real Time, EXP 5
SIP	QCI 5 (DSCP 32)	Private, EXP 4
eMBMS Data	QCI 70 (DSCP 34)	Private, EXP 4
Mobile Data – High	QCI 6 (DSCP 18)	Default, EXP 2
IR.94 Video	QCI 2 (DSCP 22)	Default, EXP 3
Mobile Data – Normal	QCI 8 (DSCP 22)	Default, EXP 1
Mission Critical PTT Voice	QCI 65 (DSCP 47)	Real Time, EXP 7
Non-Mission Critical PTT Voice	QCI 66 (DSCP 46)	Real Time, EXP 5
Mission Critical Video	TBD (QCI 2 or 70) (DSCP 46 or 34)	Real Time, EXP (TBD)
Mission Critical Signaling	QCI 69 (DSCP 47)	Real Time, EXP 7
Mission Critical Data	QCI 70 (DSCP 34)	Private, EXP 4

This table provides QCI and DSCP mapping for application types /services for FN users.

each NPSBN service and application. NPSBN provides five service packages, namely (a) Incident Uplift, (b) Platinum Level, (c) Gold Level, (d) Silver Level for primary users, and (E) Silver Level for the extended primary users. Table 10.4 summarizes the Platinum Level package. For all other services categories, QPP and DSCP value will have different assignments. The intention here is to provide readers with the concept of QPP and DSCP assignment for various services.

10.1.4.5 Supporting Up to 15 EPS Bearers per UE

Recently, 3GPP technical specifications recommend 15 EPS bearers instead of 8 bearers. To enable support for establishing 15 EPS bearers, the Evolved Packet Core (EPC) must be connected to the Evolved Universal Terrestrial Radio Access Network (E-UTRAN) to support 15 EPS bearers for FN UEs. If the UE supports 15 EPS bearers, then the UE indicates this to the MME in Non-Access Stratum (NAS) signaling as defined in TS 24.301 (46).

In the case of mobility procedures involving target nodes that do not support 15 EPS bearers, additional bearers not supported by the non-upgraded nodes should be released. The network should homogeneously support 15 EPS bearers per UE, at least per MME pool area/SGW serving area, to avoid EPS bearer deactivations and service attempts to reactivate the deactivated EPS bearers caused by UE mobility within that area. This can be achieved by defining the tracking areas accordingly.

The MME provides the UE with a Tracking Area ID (TAI) list such that it triggers the UE to perform the Tracking Area Update (TAU) procedure when the UE enters and exits the area with support for 15 EPS bearers per UE. The TAU procedure execution enables the nodes in the supporting area to identify a supporting UE when it enters the area. When exiting the area, it enables the nodes to remove bearer resources for the UE towards nodes without support for 15 EPS bearers per UE.

10.1.5 NPSBN QPP – KEY POINTS

QPP policy and profiles are important in the NPSBN to ensure that critical communications used by public safety users are prioritized, reliable, and maintained under various operating conditions. The following are the key points:

1. NPSBN QPP includes QoS and ARP. QoS includes critical elements of QCI, bit rates, packet latency, packet loss, and error, AC, etc., and

TABLE 10.4

Summary of NPSBN Service QPP and Associated DSCP

NPSBN Service Applications	QCI	Bit Rate (GBR/ Non-GBR)	Bearer (Default/ Dedicate)
Conversational Voice Service (VoLTE, Voice Mail) –IMS APN	1	GBR	Dedicated
Conversational Video – IR94 Video Calling – IMS APN	2	GBR	Dedicated
Real-Time Interactive Applications (Gaming, Monitoring, etc.)	3	GBR	Dedicated
Nonconversational Video Streaming (Buffered)	4	GBR	Default Bearer
Priority Data/Noninteractive Video – General Purpose Data APN	6	Non-GBR	Default Bearer
IPME	6	Non-GBR	Dedicated
Real-Time Text	1	GBR	Dedicated
Emergency Voice 9-1-1	1	GBR	Dedicated
Mission Critical PTT Voice (User Plane)	69	GBR	Dedicated
Non-Mission Critical PTT Voice (User Plane)	66	GBR	Default
MC Video (User Plane)	67	GBR	Dedicated
MC Data (User Plane)	70	Non-GBR	
Mobile Broadcast Multicast Service (MBMS)	70	Non-GBR	Dedicated
Wireless Priority Services	5	Non-GBR	Dedicated
IMS SIP Signaling	5	Non-GBR	Default
Interactive Service – Motion Tracking Data (Future Service)	88	Non-GBR	

ARP includes access allocation priority, PCI, and PVI. A combination of QoS and ARP policies are part of the user profile which determine the NPSBN access, priority, and retention of services of each FN user.

2. AC is assigned to each user to indicate its access priority and stored in the SIM. This classification helps manage network access, especially during high traffic situations, by determining which devices can connect to the network. Access to NPSBN is barred for the devices which have low priority.
3. FN-approved UE, NPSBN, RAN, and core have crucial roles to play in making the QPP assignment decision based on the FN user profile package. The highest QPP allocation is used for incident management. Extended primary users receive the lowest QPP among all FN users.
4. QoS and ARP play vital roles during the network congestion to authorize access to first responders based on the FN user profiles and

Latency	Packet Loss-Error Rate	AC	ARP	ARP PCI	ARP PVI	DSCP
100 ms	10^{-2}	12 or 13	3	Enabled	Disabled	46
150 ms	10^{-3}	12 or 13	3	Enabled	Enabled	22
50 ms	10^{-3}	12 or 13	3	Enabled	Enabled	
300 ms	10^{-6}	12 or 13	3	Enabled	Enabled	
300 ms	10^{-3}	12 or 13	3	Enabled	Disabled	18
300 ms	10^{-3}	12 or 13	4	Enabled	Disabled	
300 ms	10^{-3}	12 or 13	7	Enabled	Disabled	
100 ms	10^{-2}	10	7	Enabled	Disabled	
75 ms	10^{-2}	12 or 13	2,3,4 or 5	Enabled	Disabled	47
100 ms	10^{-2}	12 or 13	2	Enabled	Disabled	
100 ms	10^{-3}	12 or 13	6	Enabled	Disabled	34
200 ms	10^{-6}	12 or 13	6	Enabled	Disabled	34
200 ms	10^{-6}	12 or 13	6	Enabled	Disabled	
100 ms	10^{-2}	14	3	Enabled	Disabled	
100 ms	10^{-6}	12 or 13	3	Enabled	Disabled	32
10 ms	10^{-3}					

either block network access to non-FN users or drop them all together at the designed congestion conditions except a 9-1-1 call in progress.

5. DSCP is used for NPSBN transport quality management by defining the level of service quality that different types of data packets should receive. It helps prioritize packets for various applications, ensuring that time-sensitive services like voice or video maintain better performance across the network.
6. NPSBN operational performance is crucial for the overall capabilities of the network. These include availability, reliability, retainability, and the MTR. These NPSBN operational performance ensures NPSBN to support first responders anytime, anywhere.
7. The summary table details the various QPP and DSCP values assigned for different services and applications.

11

Satellite Networks and NPSBN Deployable

Nationwide Public Safety Broadband Network (NPSBN) uses satellite technology to ensure reliable and secure communication for public safety agencies across the United States and U.S. territories, especially during emergencies or natural disasters when terrestrial networks are down or not available at all. Satellite communication provides voice, data, and video services, enabling first responders to coordinate their efforts, share information in real time, and maintain connectivity when terrestrial infrastructure is compromised. NPSBN uses Inmarsat satellite communications and the newly developed Starlink, which is based on Low Earth Orbit (LEO). The Starlink has developed a constellation of satellites to provide high-speed internet services. There is no hands-off required to neighboring cell sites, thus X2 traffic is not involved.

First responders do not require satellite specific devices, e.g., iridium phones or any other accessories for attaching to satellites. FN-enabled devices can attach to satellite communications over a portable satellite communication system equipped with eNB/gNB Radio Access Network (RAN) having B14 spectrum. The first responders attach to eNB/gNB on the deployable like any terrestrial NPSBN RAN.

NPSBN satellite communication is called "deployable" since they can be transported and deployed anywhere if needed. NPSBN approach to deployable for the FirstNet is based on two similar but separate use cases:

- Deployables for the "Emergency Coverage Replacement," i.e., restoring services to areas normally covered by the macro B14 network after a disaster or emergency.

DOI: 10.1201/9781003557739-13

- Deployables "Coverage Extension," i.e., ad hoc deployment by public safety agencies to extend coverage to areas not normally covered by the macro RAN.

11.1 NPSBN SATELLITE NETWORK ARCHITECTURE

NPSBN satellite network comprises the following subnet units:

1. NPSBN deployable unit having B14 RAN, modem/router, satellite disc, and other necessary communication components.
2. Air link with Inmarsat over IPsec using XipLink
3. Teleports to communicate with satellites
4. NPSBN transport over AT&T Virtual Private Network (AVPN) connecting NPSBN teleport to NPSBN core over Multiprotocol Label Switching (MPLS) corporate backhaul.

The architecture illustrated in Figure 11.1 is for FN Satellite Cell on Light Truck (SAT/COLT) backhaul for two teleports (Moxee in Washington State and Castle Rock in Colorado). FirstNet requires that all the FN user traffic (S1 U) traversing through satellite transport be secured by encrypting the S1 U traffic. S1 U encryption is incorporated using a vendor product, XipLink, to safeguard FN information during the air communication between the deployable remote modem and satellite, and between the satellite to teleport. In addition, XipLink provides features of IP packet coalescing, header compression, Transmission Control Protocol (TCP) optimization/acceleration, and packet security.

As Figure 11.1 shows, NPSBN deployable has built-in RAN with B14 spectrum and a satellite's disc. First responders using the deployable service access use their User Equipment (UE) and make calls or access data services. The UE attaches to deployable eNB/gNB which is connected to remote modem and satellite disc over the cable/fiber as shown. Communication is established between satellite disc and Inmarsat. The Inmarsat transmits the signal over air to the ground-based teleport for onward transmission to FN core to perform attachment process and packet transmission. XipLink performs S1 U encryption between "remote modem" and "hub modem" providing data packet security during the air link transmission.

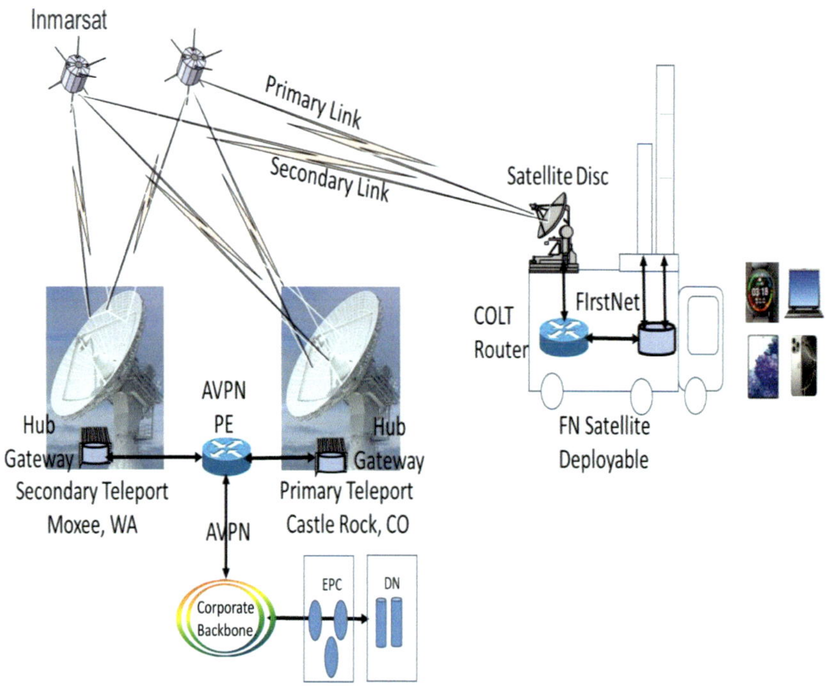

FIGURE 11.1
NPSBN SAT/COLT Deployable Logical Architecture.

NPSBN has two teleports – primary teleport located in Castle Rock and secondary teleport located in Moxee in Washington State. Currently, all deployable traffics initiating from the Continental United States, Puerto Rico, and U.S. Virgin Islands use Castle Rock teleport as primary teleport. NPSBN, deployable traffic generating from Alaska and Hawaii, use Moxee teleport as primary. Moxee also acts as back-up teleport for Continental United States, Puerto Rico, and U.S. Virgin Islands. However, the NPSBN has capabilities to configure the deployable traffics based on the prevailing conditions and demands. NPSBN also has two Mobility Management Entities (MMEs) dedicated to deployable services located in Chicago and Hazelwood, Ohio, to support FN deployable services.

11.2 NPSBN SATELLITE DEPLOYABLE SERVICE DESIGN

Currently, FN/AT&T has over 100 deployables called SAT/COLT. NPSBN also provides smaller deployable called Communication on Wheel (COW) or Compact Rapid Deployment (CRD), and a mini-CRD. SAT/COLT is owned by FN which are provided to public safety or government agencies during emergencies as needed. CRDs and mini-CRDs are mainly owned and operated by the public safety agencies. Additional details of the deployable COLT built RAN are discussed in the next section.

SAT/ broadcast FN Public Land Mobile Network (PLMN) 313-100 and has same RAN features pertaining to UE attachments, E-UTRAN Radio Access Bearer (E-RAB), Quality of Service Priority and Preemption (QPP), and interfaces with MME and SGW except the S1 encryption. In the terrestrial RAN, S1 encryption is done between eNB/gNB and Security Gateway (SGW) in the Evolved Packet Core (EPC). In case of deployables, S1 encryption is performed by XipLink between remote router and hub router. XipLink encryption applies to all traffic – control plane, user plane, and Operations, Administration, and Management (OAM). Encryption is not implemented on COLT eNB/gNB.

11.2.1 XipLink Encryption Design

Figure 11.2 illustrates high-level design of S1 encryption for SAT/COLT. As shown in the figure, S1 packets between the COLT router and hub gateway are encrypted by deploying two XipLink nodes at each end. For example, considering data packets are generated at COLT RAN by a first responder using UE, the data packets will flow to FN MME/SGW using the primary link; the XipLink deployed on the SAT/COLT will encrypt the S1 packet using the encryption algorithms. The encrypted packet is transmitted to satellite via air

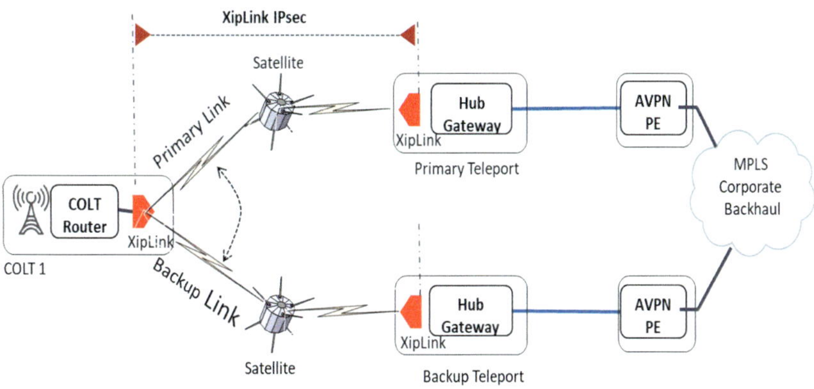

FIGURE 11.2
NPSBN SAT/COLT Service Design.

link and from the satellite to the teleport. The XipLink node deployed before the hub gateway at the primary teleport decrypts the S1 packets using IKEV2 key authentication and sends it to FN MME/SGW over AVPN. It should be noted that there is no need to encrypt the S1 packets on AVPN since the AVPN provides a secured tunnel for packet transmission. Backup link and alternate connections are designed for redundancy and serve as active routes in case any issues arise in primary link and associated route.

XipLink creates a light or outer tunnel between XipLink nodes and then creates an encrypted tunnel within the light tunnel. Encrypted tunnel carries encrypted payload to secure the data packets. In the event that the encrypted tunnel is nonoperational, the data packet will still be delivered but it would not be encrypted.

The following are the basic design requirements by XipLink to support SAT/COLT S1 encryption.

- Encapsulated Security Protocol (ESP) conforms to RFC-4303 standards. The ESP is a protocol within the IPsec for providing authentication, integrity, and confidentiality of network packets data/payload in IPv4 and IPv6 networks. ESP provides message/payload encryption and the authentication of a payload and its origin within the IPsec protocol suite.
- Tunnel mode is used between XipLink end points
- Authentication exchange is done through IKEV2 through any of the following:
  - Signatures with public key certificates (X.509) or
  - Shared secrets
- Encryption/Integrity Algorithms:
  - Encryption: AES-256 (256 bit AES-CBC, 256 bit AES-GCM with 64 bit ICV)
  - Integrity: SHA512
  - Diffie Hellman: DH group 16

Figure 11.3 provides details of packet optimizations, encrypted header, and signed ESP.

11.2.2 Transport and Tunnel Modes in IPsec

The IPsec standards define two distinct modes of IPsec operation: *transport mode* and *tunnel mode*. The modes do not affect the encoding of packets. The

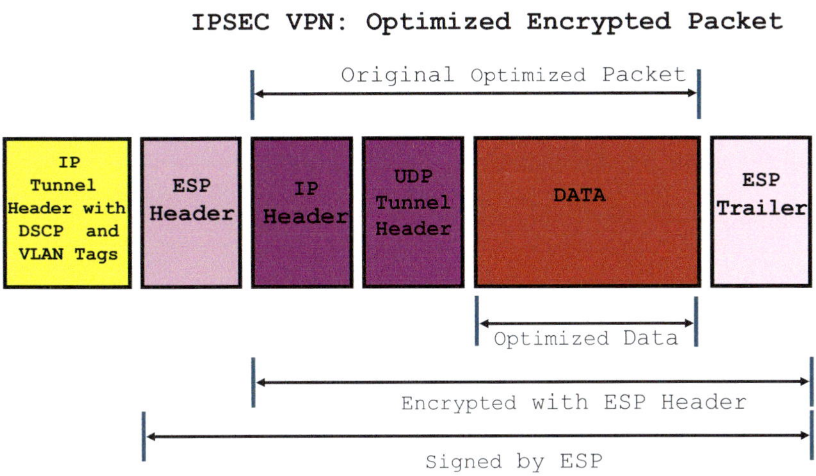

FIGURE 11.3
IPsec Protocol Structure.

packets are protected by Authentication Header (AH), ESP, or both in each mode. The modes differ in policy application when the inner packet is an IP packet, as follows:

- In transport mode, the outer header determines the IPsec policy that protects the inner IP packet.
- In tunnel mode, the inner IP packet determines the IPsec policy that protects its contents.

In transport mode, the outer header, the next header, and any ports that the next header supports are used to determine IPsec policy. In effect, IPsec can enforce different transport mode policies between two IP addresses to the granularity of a single port. For example, if the next header is TCP, which supports ports, then IPsec policy can be set for a TCP port of the outer IP address. Similarly, if the next header is an IP header, the outer header and the inner IP header can be used to determine IPsec policy.

Tunnel mode works only for IP-in-IP datagrams. In tunnel mode, IPsec policy is enforced on the contents of the inner IP datagram. Different IPsec policies can be enforced for different inner IP addresses.

11.3 NPSBN SATELLITE PORTABLE DEVICES

NPSBN provides the following deployable models to suit different deployment situations, capabilities, and transport facilities.

- SAT/COLT
- COW or CRD
- Mini-CRD

11.3.1 SAT/COLT

FN SAT/COLT deployable is completely mounted on truck and provides voice and data service when users are in range and registered on the unit with their device. These services have Quality of Service (QoS) similar to a terrestrial network. FN SAT/COLT equipped with the radio and connected to core networks provides FN functions (e.g., messaging, alerting, updating, and tracking of the deployable unit). Figures 11.4 and 11.5 show a SAT/COLT.

The COLT RAN and the disc are visible, but COLT router, modem, and other components are kept inside safely. FN SAT/COLTS are made available 24/7 to

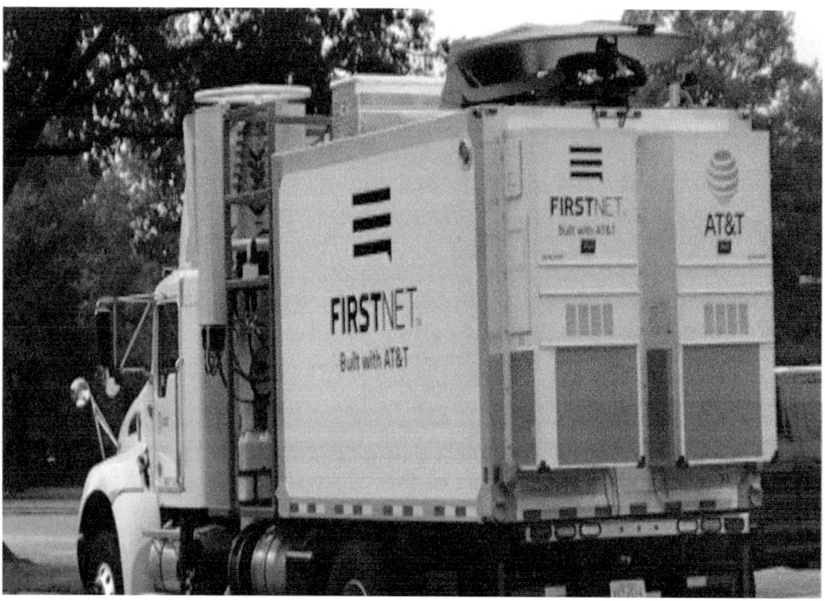

FIGURE 11.4
FN SAT/COLT.

FIGURE 11.5
FN SAT/COLT Getting Prepared for Deployment in the Field.

FN servicing agencies/government at no extra cost during planned events or emergencies. The mounted RAN on FirstNet SAT/COLT communicates via satellite and does not rely on commercial power availability. First responders attached to SAT/COLT RAN receive connectivity like conventional terrestrial cell towers (see Figure 11.6).

Currently, more than a hundred ground-based SAT/COLTs and CRDs – are in services. SAT/COLTs are vehicles with mobile cell sites that connect via satellite and do not rely on commercial power supply, while CRDs are small portable cell sites that can be brought into an area for coverage.

- Three communications vehicles – A commanding and communications vehicle for emergency deployments, planned events, and training exercises with a space for two communications personnel with multiple monitors, televisions, and charging stations, as well as a large exterior screen and speakers for briefings. It provides connectivity via Long-Term Evolution (LTE) (Band 14) and/or Wi-Fi and is able to leverage a variety of backhaul options. It is equipped with a generator that can run for multiple days before refueling and includes a lavatory, microwave, mini-refrigerator, and a sleeping bunk.
- Three airborne flying Cell on Wings – Tethered drones with larger propellers, increased payload capacity, and specialized LTE radios and power systems. Flying COWs can withstand light rain and wind speeds

FIGURE 11.6
FN SAT/COLT Deployment Briefing in Puerto Rico.

up to 25 miles per hour and reach heights of up to 400 feet, making them ideal for wildfires, mountain rescues, and other missions where the terrain previously made it difficult to maintain connectivity.

11.3.2 FN COW/CRD™

FN COW, also called CRD, are easy to transport on an SUV or a trailer and can be set up within 15–30 minutes at the site enabling the first responders to connect to their FN devices via satellite network like SAT/COLT.

The CDR provides coverage range for FN mobile devices up to two miles depending upon the prevailing conditions, e.g., land terrain, and supports all FN services and QPP including Wi-Fi range up to 500 feet and high-speed satellite internet. CRD has built-in generators for power and does not require power from the power utilities. The CRD generator can power CRD for about 60 hours before refueling is required. Figure 11.7 provides a view of FN COW/CRD.

Figure 11.8 illustrates a view of the CRD with major components listed. Similar to SAT/COLT, the CRD has its FN radio spectrum B14 and disc mounted on the CRD to communicate to satellite network.

FIGURE 11.7
FN CRD.

Figure 11.8 illustrates the CRD satellite modem, satellite controller, and the cradlepoint router for communications and a 2,200 W power generator and a power control to manage powering of the CRD components. The air conditioning and heating systems maintain the components environment during the operations.

11.3.3 FN Flying COW Drone

Recently, FN added the airborne flying COWs drone into the NPSBN services. The flying COWs are ideal for the situations where SAT/COLTs and CRDs may have logistic and location issues for deployment, e.g., in mountainous/ravine/canyon regions, wild wire, and flooded areas where the terrain made it difficult to bring SAT/COLTs and CRDs. Flying COW components include drones and a trailer for transport, equipped with a satellite dish and fiber connections. The flying COWs reach heights of up to 400 feet and maintain connectivity on the B14 spectrum, withstand light rain and wind at speeds up to 25 miles per hour.

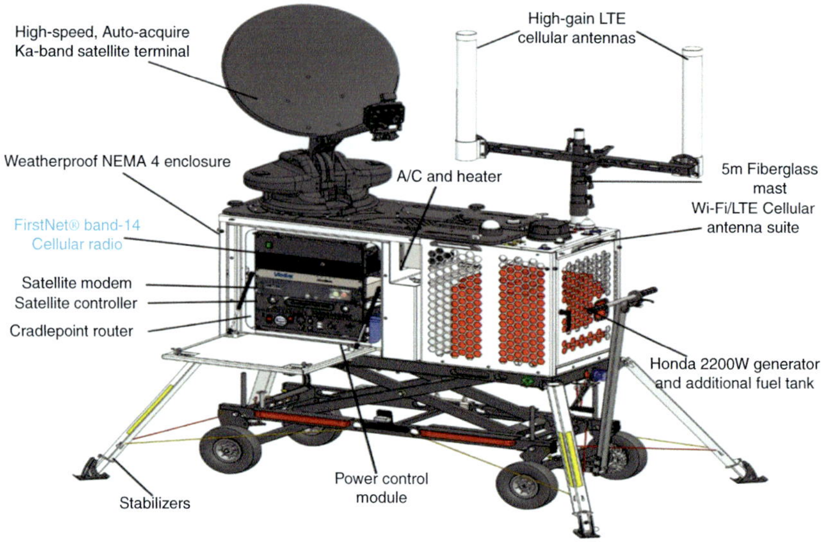

FIGURE 11.8
FN COW/CRD Components (74).

Flying COW Network Connectivity
The FirstNet flying COW is tethered to a trailer equipped with dual baseband units and satellite disc to connect either satellite or FN terrestrial cell sites. Currently, the default setting is for providing satellite connectivity like other deplorables. Satellite connectivity to and from the satellite disc mounted on towable trailer and from the satellite to teleports onward connectivity to FN EPS are similar to network connectivity used on SAT/COLTs and CRDs. XipLink provides data packet encryption to ensure data security.

11.3.4 Mini Compact Rapid (Mini-CRD) Deployable

Mini-CRD is a small portable deployable that comes in the size of a small and rugged carry-on and comprises two kits: (a) Mini-CRD case, (b) SLC+ link case, as shown in Figure 11.9. Both kits are highly portable and can simply be carried by hand, in a car, on a boat, in a plane, or in any locally available transport devices.

Starlink Communication (SLC+) is a satellite communication terminal that establishes Starlink internet connections. Starlink is a satellite internet service developed by SpaceX, which aims to provide high-speed internet access

FIGURE 11.9
FN Mini-CRD and SLC+ Communication Kits.

across the globe, particularly in underserved and rural areas. It utilizes a constellation of satellites in LEO to deliver internet connectivity, resulting in lower latency compared to traditional satellite internet services. Starlink is designed to work anywhere with a clear view of the sky and provides stable connections even in remote locations.

The Mini-CRD is connected via wired Ethernet or/and Wi-Fi to SLC+ and communicates with FN-enabled UEs. Both Mini-CRD and SLC+ contain internal batteries and power management systems but must have external power supplied by the user for sustained operations. Cases have extendable handles and can be rolled or carried. Operation is completely automatic and is controlled by two power switches on each unit. Visible lights indicate operational status. These cases can be used together or separately, depending on the scenario and use case.

FN Mini-CRD coverage area is about half a mile, and the Wi-Fi range is about 500 feet. Both kits have up to two hours of internal battery power and are easily powered by a 12V DC motor vehicle or a 120/220 V external power source. Mini-CRDs are fully tested and certified and are available in 50 states, the District of Columbia (DC), U.S. territories in the Caribbean Islands (Puerto Rico and U.S. Virgin Islands), and U.S. Pacific territories of Guam, Northern Mariana Island, and American Samoa.

11.4 NPSBN SATELLITE DEPLOYABLE QPP AND SECURITY

The FN deployables provide similar FN QPP which are available to first responders on terrestrial cell network except the packet latency due to satellite distance. On average, the packet latency is about 650 ms compared to 300 ms on terrestrial network. Since FN deployable uses B14 spectrum and attaches to FirstNet PLMN 313-100, FN QoS, Allocation and Retention Priority (ARP), Preemption Capability Indicator (PCI), Preemption Vulnerability Indicator (PVI), and priority services are available on first responder devices.

FN SAT/COLT provides S1 encryption for air link from COLT router to satellite and from satellite to teleport gateway. All FN traffic from teleports to FN EPC is over AVPN. The VPN creates a secure and encrypted connection over a less secure network, such as the internet. It enables FN users to send and receive data while maintaining privacy and anonymity by masking their IP address and encrypting their data packet payload, making it difficult for third parties to intercept or monitor their online activities. AVPN class-of-service profile is selected to map mobility applications into appropriate AVPN classes in order to maintain consistent class-of-service experience end-to-end and minimize differences with traditional backhaul QoS supported in mobility dedicated Cell Site Backhaul (CSBH) network.

In CRD and mini-CRD, cradlepoint serves as a secure networking solution and uses firewall for NPSBN connectivity. Cradlepoint enhances network security, manages bandwidth, and monitors data traffic, ensuring the protection of sensitive information and maintaining connectivity. Cradlepoint firewalls actively monitor and filter incoming and outgoing network traffic, preventing unauthorized access and blocking malicious content

11.5 NPSBN SATELLITE COMMUNICATIONS – KEY POINTS

NPSBN satellite communication services provided by the FN deployables serve crucial roles, especially in providing communication services to first responders managing emergencies in the areas where terrestrial cell sites are damaged, or the areas initially have no 4G/5G coverage. The following are the key points:

- NPSBN deployables provide satellite communication for voice, data, and video enabling first responders to coordinate their efforts, share information in real time, and maintain connectivity when terrestrial infrastructure is compromised or additional resources are required due to the presence of a large crowd, e.g., sports events like Super Bowl.
- FN provides several types of deployables to suite various conditions, locations, terrains, and types of emergencies. Currently, SAT/COLTs, CRDs, Flying COWs, Min-CRDs are included in deployable portfolio. FN continues to add new deployable devices for NPSBN services.
- Deployables provide a secure and encrypted connection for satellite communications and AVPN connections on ground transport keeping the first responders communication safe.
- Deployables deploy B14 spectrum and support FirstNet PLMN 313-100 and provide FN QPP.
- There is no hands-off required to neighboring cell sites and X2 traffic is not involved.
- Deployables support FirstNet-capable devices enabling first responders to have similar experience of terrestrial communication networks.
- Agencies subscribing to FirstNet services can request deployable support 24/7 for disasters, critical incidents, and planned events.
- The FirstNet Response Operations Group (ROG) at AT&T manages the allocations of the deployables and determines the best use of assets and vehicle platforms to address the needs based on the information provided by the requesting agencies to ROG.
- FN SAT/COLTs are provided at no cost to FirstNet subscriber agencies, including associated support costs (fuel, personnel, and satellite airtime).

12

NPSBN in the Caribbean Territories of Puerto Rico and U.S. Virgin Islands

Middle Class Tax Relief and Job Creation Act of 2012 of the U.S. Congress (1) mandated the National Public Safety Broadband Network (NPSBN) services to first responders in the U.S. Caribbean Territories of Puerto Rico (PR) and U.S. Virgin Islands (USVI). American Telephone and Telegraph Company (AT&T) had communication networks in PR and USVI but divested to Liberty Latin America (LLA) in 2022. The transaction includes network assets, including spectrum, real estate and leases, customers including wireless subscribers, and the AT&T employees working in PR and USVI to LLA. Subsequent to the divesture, FN/AT&T contractually designated LLA its vendor to support NPSBN services to first responders in PR and USVI.

12.1 PUERTO RICO

PR is an unincorporated territory of the United States located in the Caribbean, east of the Dominican Republic and west of the Virgin Islands. It has a total land area of approximately 3,515 square miles. As of 2024, the estimated population of PR is around 3.25 million people. San Juan is the capital of PR with a population of about 418,140. PR has its own constitution and a local government comprising an elected governor and legislative assembly. However, U.S. federal laws apply here, and residents are U.S. citizens but cannot vote in presidential elections. PR has its own public safety administrative infrastructures including police, fire, emergency medical service (EMS) (see Figure 12.1).

DOI: 10.1201/9781003557739-14

FIGURE 12.1
Map of PR.

12.2 U.S. VIRGIN ISLANDS

The USVI consist of three main islands: Saint Thomas, Saint John, and Saint Croix. These territories are subject to U.S. laws but have their own local governments. Saint Thomas has an area of about 32 square miles with a population of roughly 51,000. Saint John is the smallest, covering about 20 square miles and has around 4,200 residents. Saint Croix is the largest island, with an area of approximately 84 square miles and a population of about 50,000. USVIs have a locally elected governor and a legislature. Like PR, residents are U.S. citizens but do not have a vote in presidential elections. USVIs too have public safety administrative infrastructures including police, fire, EMS (see Figure 12.2).

12.3 NPSBN CARIBBEAN NETWORK ARCHITECTURE

LLA owns and manages the FCC spectrum, 4G/5G RAN, core, and transport for their subscribers in PR and USVI. As per FN/AT&T contractual arrangements, FN/AT&T has granted permission to use B14 spectrum in PR and USVI to serve FN users along with other LLA spectrum. All FN user accounts are owned and managed by FN/AT&T. The FN users include those residing in PR and USVI and visiting from the Continental United

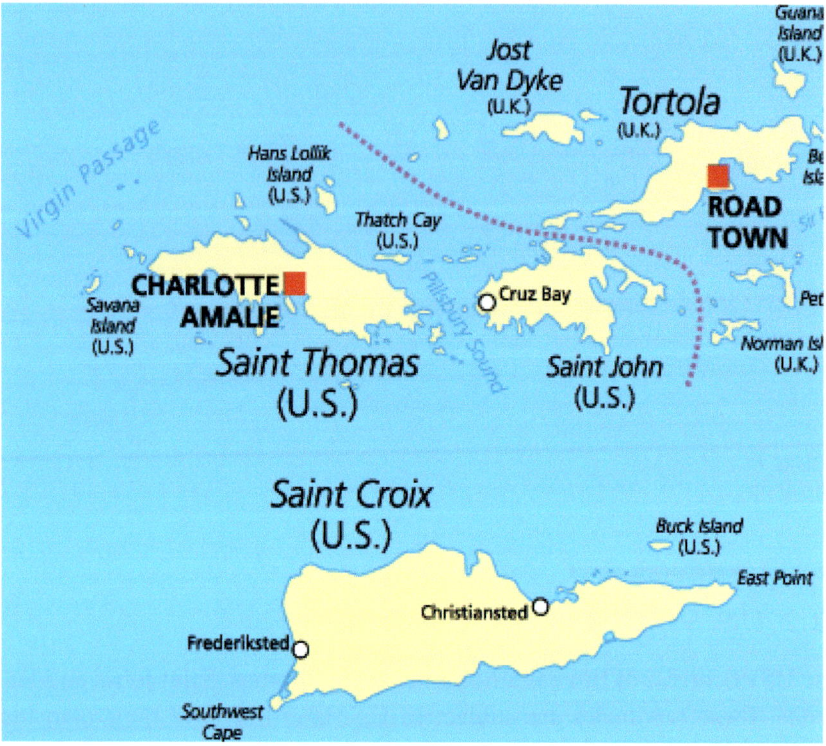

FIGURE 12.2
Map of the USVI.

States (CONUS) to PR and USVI for official or business purposes. FN customer-related services – Business Support Systems (BSS) and Operation Management Systems (OSS) – are owned and managed by FN/AT&T. LLA shares its Radio Access Networks (RANs) and backhaul between its users and the FN users but transmits FN traffic to the FN core in CONUS. FN users in PR and USVI access the NPSBN core in the U.S. Mainland, similar to how the FN users residing in United States access the FN core. LLA will support all existing and future FN services to first responders residing in PR and USVI over its shared RAN and backhaul.

Figure 12.3 illustrates the NPSBN network architecture supporting FN services in PR and USVI.

As depicted in the architecture, first responders working in PR and USVI (St. Croix, St. John, and St. Thomas) attach to LLA eNB/gNB. The signal generated by RANs located in USVI is transmitted over sea cable to LLA's Multiprotocol Label Switching (MPLS) backhaul in PR. Signals generated by

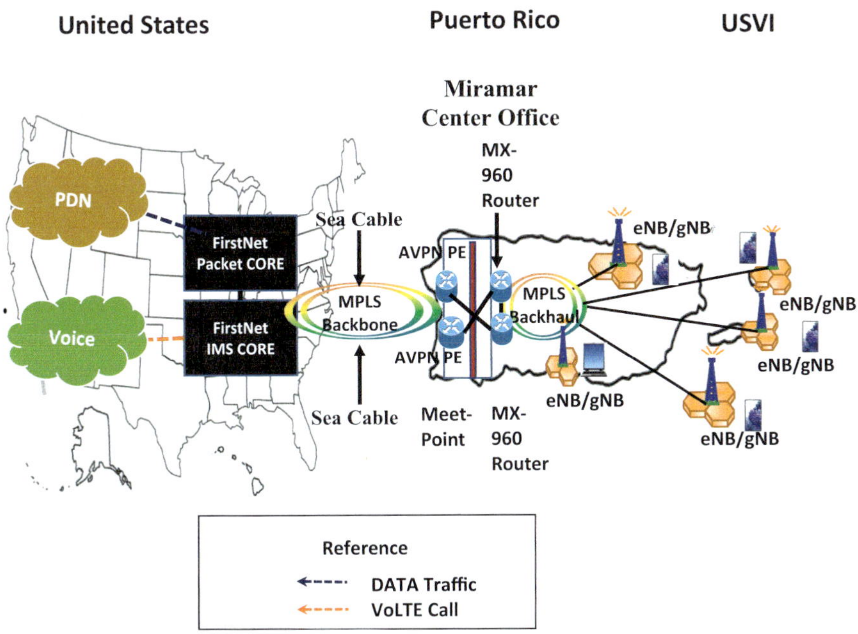

FIGURE 12.3
NPSBN Logical Architecture in Caribbean Territories of PR and USVI.

RANs located in PR are also transmitted to MPLS backhaul. The LLA MPLS backhaul transmits the signal to routers to onward transmission to the meet-point, a fiber cross-connect, in Miramar Center Office (CO). The Miramar CO is located in San Juan, PR, and owned and administered by AT&T. The meet-point transmits the signal traffic to AT&T Virtual Private Network / Provider Edge (AVPN PE) for onward transmission to FN core and FN IP Multimedia Subsystem (IMS) over AT&T MPLS backbone via sea cables.

FN Mobility Management Entity (MME) completes the attachment process and connects to FN System Architecture Evolution (SAE). The S1 U tunnel is created between FN Packet Gateway (PGW) located in the U.S. Mainland and the eNB/gNB located in PR and USVI. Then data packet or VoLTE connection is established.

12.4 NPSBN CARIBBEAN NETWORK DESIGN

LLA owns the RAN and transport in PR and USVI and facilitates first responders to share these resources.

12.4.1 LLA's RAN Design for FirstNet Users

LLA uses the same RAN vendors (Ericsson and Nokia) as AT&T, with FN RAN features and release that are on par with AT&T. AT&T and LLA teams work closely to certify that the LLA's RAN features and releases to ensure that FN users in PR/USVI do not face any issues accessing LLA's RAN. LLA's eNBs will encrypt the S1 traffic and send it to the FN core in Mainland as shown in architecture. LLA also uses small cells which will support FN services similar to eNB/gNB. LLA RAN uses separate VLANs for FN S1, operations, administration, and maintenance (OAM), and for connectivity between LLA's laboratory and FN laboratory. Interlab connectivity is designed to align new FN features and services in PR/USVI for verification and certification of the FN services in LLA before launching.

12.4.2 LLA Transport Design for FirstNet Users

LLA's mobility transport infrastructure collects and brings FN traffic from various eNB/gNB to MPLS backhaul. LLA works with AT&T to ensure that its transport network supports the prescribed Quality of Support (QoS)/Differentiated Services Code Points (DSCP) mapping and other applicable specifications. LLA is responsible for developing and testing system-to-system connectivity for its network nodes, EMS, and OSS/BSS. LLA will bring the FN traffic to the meet-point in Miramar 901 CO, which is the demarcation point between LLA and AT&T networks. AT&T owns and manages meet-point and is responsible for traffic transmission to the FN core in the Mainland using two 10 Gigs undersea cables. AVPN routers will connect to AT&T IXC zone(s) and FN core via CBB (MPLS) network. The are three AVPN circuits from LLA to the U.S. Mainland – one for S1 U/C to the FN core, a second for EMS feed to OSS, and a third for the FN/AT&T lab to LLA's lab connectivity. The fiber/copper cables between routers and undersea cables are physically redundant and have failover capability. LLA feeds its EMS data to AT&T's CANOPI and SMART platforms for the inventory of LLA's cell towers and provisioning of FN services to FN users residing in PR/USVI. Figure 12.4 illustrates a high-level VLAN design between LLA and FN/AT&T networks in the Mainland.

As illustrated in the figure, FN traffic generated in PR and USVI flows to CONUS via undersea cables. There are two 10 Gigs undersea cables which carry active payload to separate COs in Florida. Three Virtual Local Area Networks (VLANs) are part of the payload. VLAN1 connects the FN core

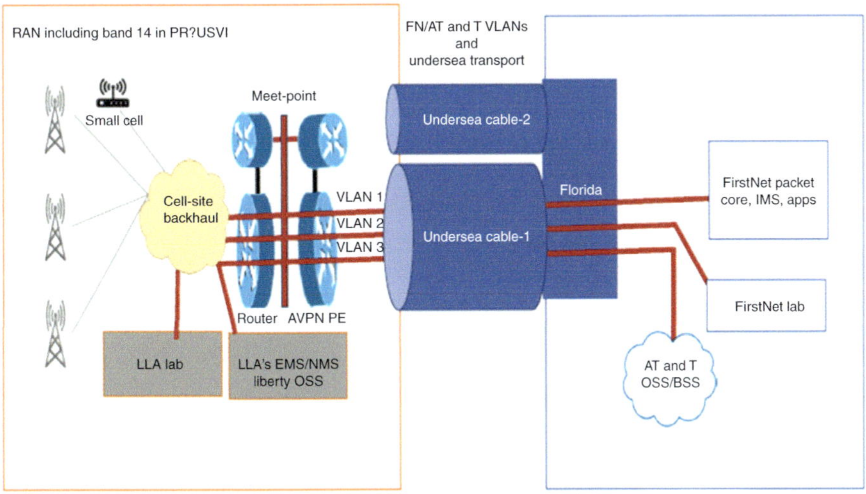

FIGURE 12.4
NPSBN Transport Design in PR and USVI.

and carries all FN traffic, VLAN2 connects LLA's lab to FN lab and carries lab-related traffic. VLAN 3 connects LLA's EMS/NMS/OSS to AT&T OSS/BSS. The transport is designed to provide local and geo-redundancy. In case an undersea cable is cut or nonoperational, the AVPN PE will send the traffic to other undersea cables via locally redundant AVPN PE. In an unforeseen situation, if both undersea cables are cut or not operational, the FN users will roam in PR and USVI. Thus, FN users are not isolated at any time due to transport-related issues and carry their public safety work 24×7.

12.5 NPSBN SERVICES IN PR AND USVI

All FirstNet services available to FN users in CONUS are available in PR and USVI. However, Wireless Emergency Alert (WEA) will be delivered to FN users in PR and USVI by LLA rather FN/AT&T. The SAT/COM and CRD serve the FN users in PR and USVI as well. Since SAT/COLT and CRD have their own RAN, the FN users attach directly to deployables and connect to FN core via teleports in CONUS. FN users in PR and USVI need their FN-enabled devices to connect to FN deployables similar to FN users in CONUS (see Figures 12.5 and 15.6).

FIGURE 12.5
FN SAT/COLT Deployment in PR.

Roaming services in PR and USVI are available to FN users. The only time FN users in PR and USVI roam is when both undersea cables are cut, as stated earlier. Emergency 9-1-1 services will be available similar to CONUS; however, the routing will be sent to PSAP in PR and USVI.

12.6 NPSBN QOS, PERFORMANCE, AND SECURITY IN PR AND USVI

LLA supports all FN QPP which are available to FN users in CONUS. In addition, LLA's RAN designs have S1 encryption and other FN security

FIGURE 12.6
CRD Demonstration in USVI.

requirement features in place. LLA provides FN service QoS, priority, and preemption (QPP) and FN operational reports to AT&T to incorporate into the final reports which are sent on a regular basis to FN authority. Operational issues and trouble resolutions will be shared between LLA and AT&T teams. RAN and transport-related issues up to the network demarcation at the meet-point are managed by LLA. AT&T is responsible for any issues from the demarcation points, including undersea cables and transport subnetwork in CONUS and FN core.

LLA provides network coverage and NPSBN key performance indicators (Net P1, P2, P3, and P4) on a regular basis to FN/AT&T which are reviewed by joint teams. Also, LLA reports on any major events, hardware/software change, and upgrade management reports.

12.7 NPSBN IN CARIBBEAN TERRITORIES – KEY POINTS

LLA plays vital roles in assisting FN/AT&T to provide NPSBN services in PR and USVI of St Croix, St. John, and St. Thomas. FN/AT&T and LLA teams have developed an extremely cohesive and comprehensive teamwork resulting in effective and efficient public safety service availability in the U.S. Caribbean territories. The following are the key points:

- LLA is NPSBN service provider in PR and USVI and owns the spectrum, RAN, core, and mobility infrastructure.
- FN/AT&T has contractual vendor relationship with LLA and granted use of B14 spectrum. LLA shares eNB/gNB and cell site backhaul between FN users and their own subscribers in PR and USVI. LLA sends the FN user traffic to FN core in CONUS, and LLA subscribers will access the LLA core in PR, not the FN core.
- Meet-point in Miramar 901 CO, San Juan in PR, is the network demarcation point between LLA and FN/AT&T network. AT&T owns and manages Miramar 901 CO.
- Two 10 Gigs undersea cables carry the FN traffic from PR to CONUS.
- LLA uses Ericsson and Nokia RAN and align RAN features and software releases with FN/ATT's RAN team ensuring to support all FN services, and FN QPP at par with FN services in CONUS. AT&T and LLA teams coordinate the features and releases schedule. LLA also partners with Samsung, especially for for 5G networs.
- LLA's mobility lab in San Juan has connectivity to interface and shares information with FN lab in Redmond, Seattle. FN labs certify the features and RAN software releases before the live launch.
- LLA provides its ENM/EMS information to AT&T OSS/BSS for inventory and provisioning alignments.
- FN deployable in PR and USVI has no interaction with LLA RAN and transport. FN users having FN-enabled devices attach to deployable RAN similar to FN users in CONOUS.
- LLA and AT&T have operations teams who meet and resolve issues related to network.

13

NPSBN in Pacific Territories

The Nationwide Public Safety Broadband Network (NPSBN) program is part of a federal initiative in the United States to create a reliable and secure communications network specifically for public safety responders, including police, fire, and emergency medical services (EMS). In the U.S. Pacific Islands, the program aims to improve communication capabilities and interoperability among first responders, ensuring that the first responders respond and manage emergencies effectively.

The U.S. territories in the Pacific region comprises the following three islands which are also collectively termed "3PT" (see Figure 13.1).

1. Guam
2. Commonwealth of Northern Mariana Island (CNMI) - Saipan
3. American Samoa

13.1 GUAM

Guam Island is located in the Northern Pacific Ocean having latitude and longitude of 13.4443° N and 144.793732° E, respectively. Guam is 30.17 miles (48.55 km) long and 4–12 miles (6–19 km) wide. It has an area of 212 square miles (549 sq. km) and a population of approximately 168,000 as of the most recent estimates (see Figure 31.2). The government of Guam is an unincorporated territory of the United States, which means that while it has its own local government, the U.S. federal government retains sovereignty and jurisdiction over certain areas. Guam's local government is elected and includes an

DOI: 10.1201/9781003557739-15

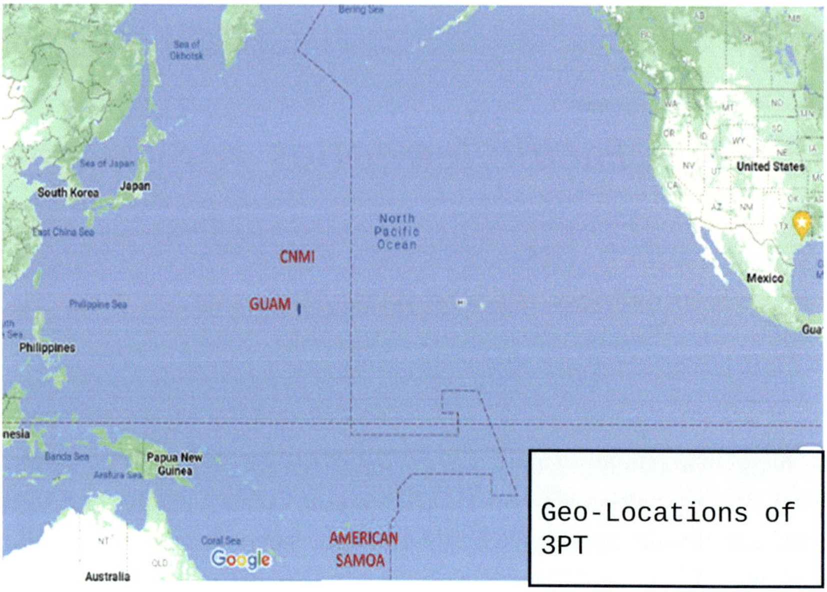

FIGURE 13.1
U.S. Territories in the Pacific Region.

FIGURE 13.2
Guam at a Glance.

elected governor, legislature, and various local offices including public safety agencies, e.g., police, fire, EMS, etc. Guam also has a large U.S. defense base.

13.2 CNMI – Saipan Island

Saipan Island is located in the northern part of the Mariana Islands in the western Pacific Ocean. It is the largest island in the CNMI and lies about 120 miles (193 km) north of Guam. Saipan latitude is 15.1850° N and longitude is 145.7467° E, respectively. The latest estimate of the population of the CNMI is approximately 50,000 and the population of Saipan is approximately 48,000.

The government of the CNMI operates under a local constitution and is a self-governing territory of the United States. It has three branches: the Executive branch, led by the Governor; the Legislative branch, consisting of the Senate and the House of Representatives; and the Judicial branch, which includes local courts. Federal agencies that operate in the CNMI include the Department of the Interior, Federal Bureau of Prisons, and the Department of Homeland Security, among others, overseeing various federal functions and support. Figure 13.3 illustrates the Saipan Island territorial map and the local towns and roads.

FIGURE 13.3
Map of the Saipan Island.

13.3 AMERICAN SAMOA

American Samoa archipelago is an unincorporated territory of the United States. It is located in the South Pacific Ocean having the latitude of 4.2710° S. and longitude of 170.1322°W., respectively. American Samoa lies about 1,600 miles (2,600 km) northeast of New Zealand and 2,200 miles (3,500 km) southwest of the U.S. state of Hawaii. It includes several islands, with the largest being Tutuila. Its current population is about 57,881, who are all U.S. citizens but are not eligible for voting for the U.S. President. American Samoa is administered by the locally elected government, the elected governor being the head of the Executive branch. American Samoa government has its own public safety agencies and telecom service providers who serve local residents.

Figure 13.4 provides a geographical map of Tutuila which is the largest island in the American Samoa archipelago.

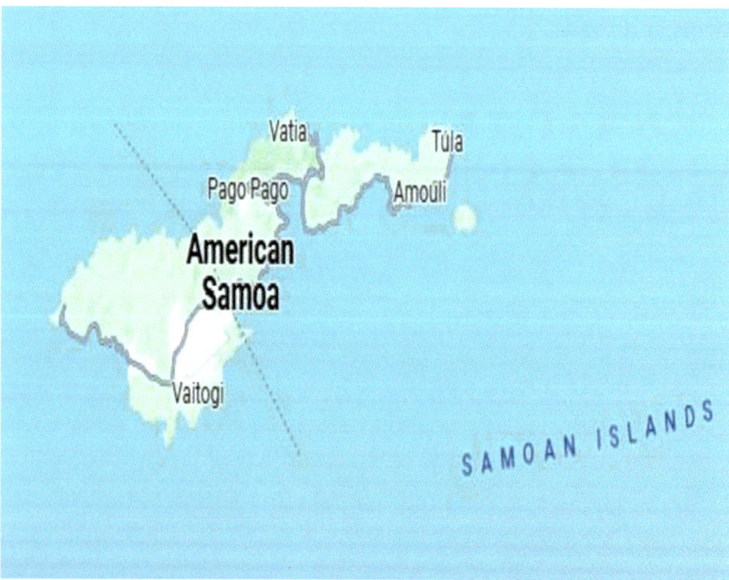

FIGURE 13.4
American Samoa – Tutuila Island Map.

13.4 FEDERAL COMMUNICATIONS COMMISSION SPECTRUM ALLOCATION IN 3PT

The Federal Communications Commission (FCC) has licensed spectrum bands to serve 3PT. The Do Communications over the Mobile Network (DOCOMO) Pacific, owned by Nippon Telephone and Telegraph (NTT) of Japan, is a major mobile service provider in Guam and CNMI. DOCOMO is also termed DPAC (DOCOMO Pacific). DPAC is a service provider in GUAM as well as CNMI. The American Samoa Telecommunications Authority (A STCA) is a telecommunications service provider in American Samoa, owned by the American Samoa Government and licensed by the FCC. Currently, both DPAC and ASTCA have Long-Term Evolution (LTE) spectrum licenses including permission to use B14 to their customers. Both DPAC and ASTCA are running feasibility trials in their networks to provide 5G services. DPAC Radio Access Network (RAN) vendor is Nokia and ASTCA has Ericsson Nokia and Ericsson also provide RAN equipment to American Telephone and Telegraph Company (AT&T). Having common RAN vendors makes it technically easier to align the NPSBN features and services compatible in Contiguous United States (CONUS) and 3PT. Table 13.1 provides the list of the spectrum bands licensed to DPAC and ASTAC in 3PT.

TABLE 13.1

FCC Spectrum Licenses in 3PT

DPAC – GUAM	Technology	BAND	Frequency
	LTE	66	10MHz
	LTE	12	5MHz
	LTE	14	10MHz
	LTE	71	20MHz
	LTE	41	10MHz
	NR	41	10MHz
DPAC- CNMI (SPN, TIN, ROT)	LTE	66	10MHz
	LTE	12	5MHz
	LTE	14	10MHz
	LTE	41	10MHz
	NR	41	10MHz
ASTCA – American Samoa	LTE	2	10MHz
	LTE	12	10MHz
	LTE	14	10MHz
	NR	12	10MHz

13.5 NPSBN PACIFIC ISLANDS NETWORK ARCHITECTURE

Due to the location of the 3PT, it is rather a disadvantageous endeavor to connect the FN users residing in 3PT to the FirstNet (FN) core in CONUS due to latency and other operational issues. FN/AT&T made a strategic decision to support NPSBN service locally by having strategic partnership with the local service providers and enhancing their mobility network infrastructure to provide NPSBN services to FN users residing in 3PT. FN/AT&T has partnered with DPAC to support NPSBN services in Guam and CNMI and ASTCA in American Samoa. In addition, DPAC and ASTA were granted permission to use B14 spectrum in their respective coverage regions.

FN/AT&T developed a common NPSBN service architecture for the entire 3PT focusing on SIM specifications for 3PT, FN user access to DPAC and ASTCA networks, NPSBN services in 3PT, Quality of Service (QoS), performance reporting, network security, roaming, etc. Figure 13.5 illustrates the common 3PT NPSBN network architecture to support FN users in 3PT. As illustrated in the figure, both DPAC and ASTCA comply with the 3GPP specifications to support LTE/5G network. FN users residing in 3PT coverage areas or visiting from the United States having FN-certified devices attach to 3PT RAN and avail Voice Over LTE (VoLTE), messaging, and data services as per their user profiles.

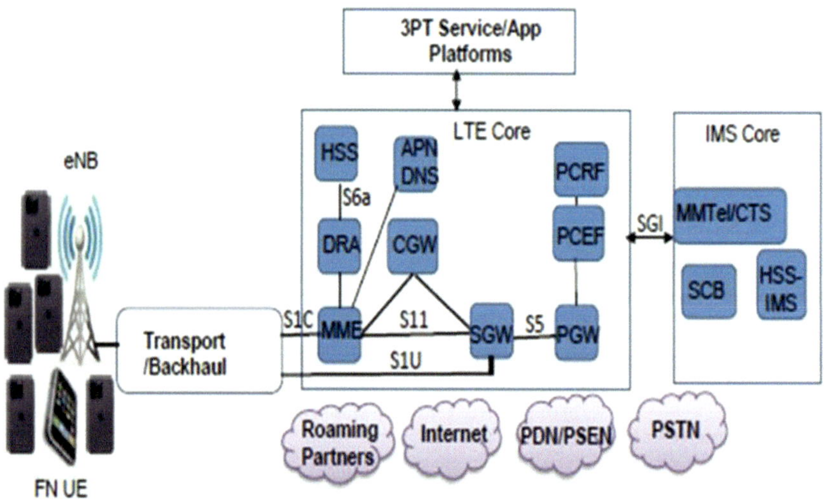

FIGURE 13.5
3PT NPSBN Service Network Architecture.

13.6 3PT NPSBN SERVICES DESIGN

It is important that FN users residing in 3PT attach to local RAN and core in 3PT coverage areas as well RAN and FN core in CONUS during their official visit to the United States. Also, FN users residing in CONUS but visiting 3PT for official work should be able to access 3PT RAN as a roamer. This network access flexibility was achieved with a specific SIM design for 3PT.

In general, a physical mobile SIM card has one International Mobile Subscriber Identity (IMSI) and one Mobile Station International Subscriber Directory Number (MSISDN). MSISDN is associated with only IMSI. MSISDN is essentially the same as a phone number. It is the number assigned to a specific mobile phone subscriber, allowing them to be reached via the mobile network. However, the SIM card designed for 3PT has two IMSIs also called dual persona SIM. One IMSI is associated with one MSISDN and the second IMSI with the second MSISDN. However, depending upon the location of the FN user, only one IMSI/ISDN is exposed, and the other IMSI/ISDN remains dormant. This dual persona SIM uses different Public Land Mobile Network (PLMN) for 3PT.

Guam/CNMI:
One SIM with two IMSIs:

1. One IMSI for Guam and CNMI (313-110)
2. One for outside Guam/CNMI (313-100)
   FN users residing in GUAM/CNMI will use IMSI 313-110 as homer in DPAC network and IMSI 313-100 will be hidden. The same user while travelling in the United States will use 313-100 to connect as homer to NPSBN services in the U.S. Mainland not using 313-110. However, if the user is travelling to American Samoa or any other location beyond Guam/CNMI and the United States, it would be treated as a roamer and would roam using 313-100 IMSI on AT&T roaming partner's networks.

For American Samoa:
One SIM with two IMSIs:

1. One IMSI for ASTCA (313-120)
2. One for outside Guam/CNMI (313-100)

13.7 NPSBN SERVICES IN PACIFIC ISLANDS

The essential NPSBN features and services available to first responders are listed here:

- VoLTE
- Data, SMS/Messaging
- 9-1-1 calling
- Lawful intercept (for authorized users)
- Roaming
- Device management
- Access to commercial LTE spectrum as well as B14 spectrum, wherever deployed

Currently, Wi-Fi calling is not available in 3PT. FN/AT&T is actively working with DPAC and ASTCA to support Wi-Fi calling and other features in the near future.

Deployables are available in Guam, CNMI, and American Samoa. The Department of Homeland Security in Guam and ASTCA have deployables to deploy as and when needed.

13.8 NPSBN QUALITY OF SERVICE, PRIORITY, AND PREEMPTION IN THE PACIFIC ISLANDS

DPAC and American Samoa provide FN quality of service, priority, and preemption (QPP) which are summarized in this section.

13.8.1 NPSBN QPP in the Pacific Islands

1. **FN QPP/Priority and Preemption**
   FN Primary Users – QCI: 6, ACB: 12, ARP: 3, PCI: Yes, PVI: No
   FN Extended Users – QCI: 6, ACB: 13, ARP: 5, PCI: N0, PVI: Yes

13.8.2 NPSBN Performance in the Pacific Islands

DPAC and ASTCA provide their network coverage and NPSBN key performance indicators (Net P1, P2, P3, and P4) on a regular basis to FN/

AT&T which are reviewed by joint teams. Also, 3PT partners report any major events, hardware/software change, and upgrade management reports.

13.8.3 NPSBN Security in the Pacific Islands

Network security is vital to protect the integrity, confidentiality, and unauthorized access, misuse, or destruction of networked systems and data from cyber threats, ensuring that sensitive information is safeguarded against various types of attacks. Also, DPAC and ASTCA have coordinated their security measures with FN/AT&T teams. 3PT comply with the various security specifications provided by AT&T. 3PT provides security reports to FN/AT&T at a regular interval. It also has the network hardening required by FN/AT&T in place.

13.9 NPSBN PACIFIC ISLANDS – KEY POINTS

The U.S. territories in the Pacific Ocean – Guam, CNMI, and American Samoa, constantly face natural disasters primarily due to their geographic locations which are susceptible to typhoons, earthquakes, and volcanic activities. These islands are also situated near tectonic plate boundaries, making them vulnerable to seismic activities, while tropical climate contributes to frequent typhoons, especially during the Pacific typhoon season. FN made a vital move to include 3PT within the realm of NPSBN to support first responders in the islands. Instead of being isolated and away from CONUS, 3PT has made significant achievements in supporting NPSBN service.

1. DPAC is the FN/AT&T partner to support NPSBN services in Guam and CNMI.
2. ASTCA is the FN/AT&T partner to support NPSBN services in American Samoa.
3. Collectively, the U.S. territories in Pacific Ocean are termed "3PT" for brevity.
4. FN/AT&T granted permission to 3PT to use B14 spectrum which has been assigned for public safety by FCC and licensed to FNA for NPSBN.

5. FN users residing in 3PT have "dual persona" SIM having two IMSIs and two MSISDN. FN users in DPAC have 313-110 and 313-100 IMSIs. FN users residing in Guam/CNMI use IMSI 313-110, but while travelling to CONUS use IMSI 313-100. Similarly, FN users residing in American Samoa use IMSI 313-120, but when travelling to CONUS use IMSI 313-100.
6. 3PT supports NPSBN services and QPP, access control, and traffic management based on user profiles.
7. FN deployables are deployed and operational in 3PT to provide communications services when the terrestrial cells are down or there is no coverage at all.
8. 3PT and FN/AT&T teams work closely and meet at regular intervals to share information, review operational reports, and plan for new features/services (see Figures 13.6, 13.7, 13.8 and 13.9).

See Figures 13.6–13.9.

FIGURE 13.6
A CRD Deployed in Guam.

FIGURE 13.7
Author is with a First Responder in Saipan Island (CNMI).

FIGURE 13.8
Author with Governor (Right) and Deputy Governor of American Samoa.

FIGURE 13.9
FN/AT&T Team at Governors' Residence in American Samoa.

14

Rural Partners Networks and NPSBN Interoperability

Middle Class Tax Relief and Job Recovery Act of 2012 authorized FirstNet Authority (FNA) to establish a nationwide Public Safety Broadband Network (PSBN) based on a single, national network to build a dedicated, reliable wireless network for first responders in all U.S. states and territories, including rural communities and tribal nations. FNA partnered with American Telephone and Telegraph Company (AT&T) to develop, deploy, and operate NPSBN as mandated by the U.S. Congress. AT&T covers most of the places in the United States, but has gaps in the rural areas including Iowa, Minnesota, Illinois, Wisconsin, and Texas. These areas are served by smaller service providers, including Cover, Panhandle, and Bug Tussel. AT&T partnered with these small service providers so that FirstNet users can get access to their broadband radio network for serving in their coverage areas.

FN users on rural partner networks are connecting to rural partners' Long-Term Evolution (LTE) Radio Access Network (RAN) and backhaul, similar to how FN users connect to FN/AT&T RAN. These users are not roaming on the rural partners' networks and do not need any modification or enhancements on FN devices to access the rural partner's network.

14.1 RURAL PARTNER NETWORK INFRASTRUCTURE

The rural partners have their own broadband network infrastructure including RAN, transport backhaul, Evolved Packet Core (EPC), and operational services to provide LTE services to their customers. As a broadband service providers, they are responsible for maintaining their network infrastructure, providing customer supports, and billing services to their customers.

DOI: 10.1201/9781003557739-16

14.2 FN USERS ON RURAL PARTNER NETWORKS

FirstNet/AT&T partnership provids the following provision to these rural partners:

- License to use B14 spectrum in their service areas
- Network Element Management, Operational Support Systems (OSS), and Business Support Systems (BSS) are provided by AT&T-supporting FN users on B14 RAN
- FN users attach to rural partners' RAN but all the FN traffic flows to FN/AT&T NPSBN.
- FN users' accounts are owned and managed by FN/AT&T and operational support services are provided by FN/AT&T.

14.3 INTEROPERABILITY BETWEEN RURAL PARTNER NETWORK AND NPSBN

Rural partner's LTE broadband architecture and design are based on 3GPP specifications; hence they are similar to AT&T RAN network infrastructure. Figure 14.1 provides a simplified view of FN/AT&T NPSBN and rural partners' LTE RAN and backhaul transport. FN traffic from rural partner's

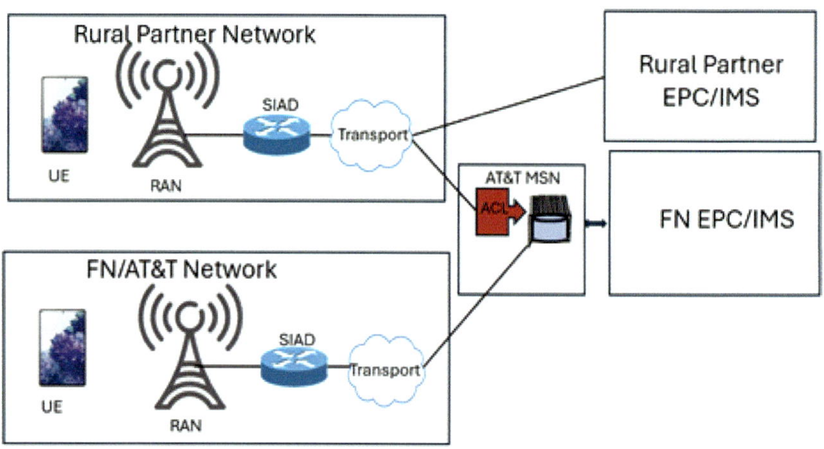

FIGURE 14.1
Interoperability between Rural Partner's Network and NPSBN.

network goes to AT&T MSN and it is then transmitted to FN EPC/IMS. Rural partners send their users' data directly to their own EPC/IMS.

FN users' traffic originating from rural partner's network is screened by Access Control List (ACL) before reaches to AT&T Mobile Station Node (MSN). The ACL list provides cybersecurity checkpoints for any traffic entering from a third-party network into AT&T MSN and allows only the traffic that has appropriate IP address.

14.4 RURAL PARTNER NETWORK INTEROPERABILITY – KEY POINTS

- Rural partners provide valuable services to FN users in their network serving areas to access NPSBN. The FN users attach to NPSBN like any other FN/AT&T RAN without necessitating any outbound roaming.
- FN/AT&T partnership enables the rural partners to use B14 spectrum and receive technical support from FN/AT&T.
- All FN users attaching the rural partner's RAN are owned and managed by AT&T. AT&T provides billing, operational supports, and customer services to all FN users.

15

NPSBN Devices

FirstNet supports a wide variety of approved FirstNet Trusted™ devices, including Android®, Apple® iOS smartphones, feature phones, and data-only devices, such as tablets, laptops, air cards, notebooks, and mobile hotspot devices. FN/AT&T updates the FN device list as and when new devices are available in the market and approved for first responders. The readers are encouraged to look the official website of the FirstNet for most recent list of the FN-approved devices.

15.1 SMART PHONES

FN offers the following smartphones to first responders

- Apple iPhones: iPhones with FN SIM (313-100) are the FN-approved devices which includes the latest iPhone 16.
- Samsung phones: FN -approved Samsung devices includes Galaxy series – Galaxy A, Galaxy S, Galaxy X, and Galaxy Z, which are folding phones (e.g., Samsung Galaxy Z Fold 5, Samsung Galaxy Z Fold 6)
- Google phones: FN-approved Google phones include Google Pixel Fold, Google Pixel 7, and the latest Pixel phones
- Motorola phones: FN-approved Motorola phones include Motorola Razr folding phones, Motorola Moto G Stylus
- Sonim phones: FN-approved Sonim phones include XP series (e.g., XP10, XP3 Plus, XP5 Plus)
- Zebra phones: Zebra phones are new addition of the FN-approved devices which include Zebra EC55, Zebra MC2700, Zebra TC series (e.g., TC57, TC26, TC77, TC78)
- Siyata SD 7 phones

DOI: 10.1201/9781003557739-17

15.2 SMART WATCHES

Number syncing of the smartphones and watches (wearables) enhances first responders' experience as they can use either device for VoLTE calls to send/ receive text messages and data. FN users can also use wearables in place of their FN-enabled smartphones in case the phone is not with them or the situation does not warrant using one. The watches have FN SIMs (313-100 PLMN) and their own cellular telephone numbers. The FN-approved watches are fully integrated and certified by FN to attach to NPSBN and support FN services. The list here provides the FN-approved major smart watches:

- Apple watch series 8, 9, 10, and Apple watch ultra
- Samsung Galaxy watches 5, 6, 7
- Google Pixel watches 2 and 3

15.3 MCPT DEVICES

FN-approved Mission Critical Push to Talk (MCPT) devices are critical for first responders in emergency situations. These devices, except L3Harris devices, have downloadable clients to launch MCPT services and are approved by FN.

- Apple iPhones 12 and the latest ones
- Samsung phones GS21, GS22, GS23, and the latest ones
- L3Harris XL Onboard Mobile and XL Extreme Portable have embedded clients and also have two-way radio features

16

The NPSBN Security and Data Privacy

Network security involves measures and protocols designed to protect the integrity, confidentiality, and availability of the networks and their data. It encompasses hardware and software technologies, policies, and practices to safeguard against cyber threats, unauthorized access, and data breaches. Effective network security ensures safe operations for networked devices and prevents malicious attacks.

Network privacy refers to the protection of personal and sensitive information transmitted over a network. It encompasses various technologies and practices designed to safeguard this data from unauthorized access, interception, or misuse. This includes the use of encryption, secure protocols, and privacy policies that govern data handling.

Network privacy and security are complementary concepts but are distinct in their focus. Network security primarily aims to protect the network infrastructure from unauthorized access and attacks, ensuring the integrity and availability of data. Network privacy, on the contrary, focuses on safeguarding personal and sensitive information shared over the network from being accessed or disclosed without consent. Together, they enhance the overall safety and trustworthiness of the NPSBN, but they address different aspects of data protection. The following security risks are most common:

- API exposure attacks
- Botnet/Malware abuse attacks
- Distributed Denial of Service (DDoS) against public IP attacks
- Domain Name System (DNS) attacks
- Flooding the interface attacks

DOI: 10.1201/9781003557739-18

- Internet facing server/application attacks
- Malware delivery attacks
- Protocol manipulation attacks
- Spoofing attacks
- Unauthorized user access attacks

16.1 NPSBN PILLAR OF SECURITY AND DATA PRIVACY

National Public Safety Broadband Network (NPSBN) pillars of security and data privacy focus on protecting user data, ensuring the integrity and confidentiality of communications, and safeguarding against unauthorized access. Key features include strong encryption techniques, user identity protection, secure signaling procedures, and continuous authentication mechanisms to prevent threats and enhance trust. It is of the utmost importance to keep FirstNet user communications safe from ID theft, hacking, intrusion, and unauthorized transfer of data from the network, which often involves stealing sensitive information and can be part of a broader cyberattack or security breach.

The NPSBN pillars of security involve architecting the network security design, installing, and maintaining specific security measures, constantly monitoring, and maintaining the network access activities over communication network at all levels and taking the operational measure to resolve the security issues. The NPSBN security pillars also include:

- Segregation of different traffics
- Security control measures
- Automated security protection logics and devices
- Monitoring various types of traffics
- Testing the suspected traffics
- Operational response readiness to combat security and data privacy threat
- Security standards and NPSBN security

NPSBN security encompasses a multilayered architecture involving security measures at each layer. A multilayered security architecture isolates the security and data privacy of the various NPSBN subsystems and components.

16.2 NPSBN SECURITY AND DATA PRIVACY ARCHITECTURE

Figure 16.1 illustrates the NPSBN multilayered security architecture. This architecture provides a comprehensive approach to cybersecurity that implements multiple layers of security controls and measures to protect an organization's data and assets. This strategy relies on various defensive mechanisms, such as firewalls, intrusion detection systems, user authentication protocols, and employee training, to create a robust defense against diverse threats, ensuring that if one layer is breached, others continue to provide protection.

To that end, the solution should consider the following areas:

- Environmental and physical layer security
- Network layer security
- User level security
- Information and data security

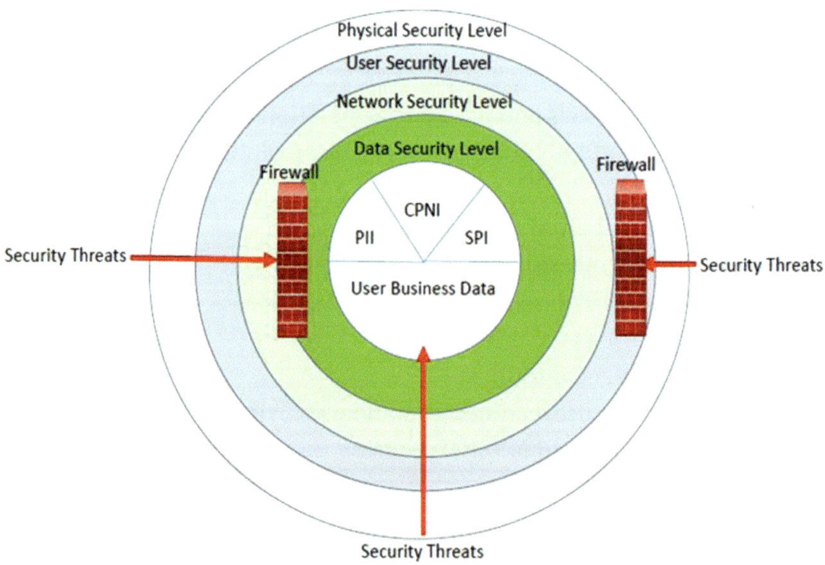

FIGURE 16.1
NPSBN Multilayered Security Architecture.

16.3 ENVIRONMENTAL AND PHYSICAL SECURITY

Physical security involves safeguarding the physical components of the network infrastructure from unauthorized access, damage, or interference. This includes securing FN devices, base stations, antennas, and associated equipment to prevent tampering, theft, or environmental hazards, thereby ensuring the integrity and availability of the network services.

In NPSBN, physical security measures are essential to protect network infrastructure and prevent unauthorized access. The types of physical security in place include:

1. *Access Control:* Restricting entry to network facilities through identification systems, security personnel, and electronic access systems.
2. *Surveillance and Monitoring:* Utilizing CCTV cameras and alarm systems to monitor facilities and detect unauthorized activities.
3. *Environmental Protection:* Implementing measures against natural disasters (floods, earthquakes) and vandalism, including reinforced structures and climate control.
4. *Hardware Security:* Protecting physical devices, such as routers and switches, through tamper-proof enclosures and physical locks.
5. *Secured Transmission Facilities:* Ensuring that transmission points (e.g., base station sites) have been secured physically against attacks or unauthorized access.

It is crucial to integrate these measures into an overall security strategy to effectively safeguard the Long-Term Evolution (LTE) network.

16.4 FN USER LEVEL SECURITY

The user level security involves two aspects of security: (a) devices security and (b) access to network and applications using the FN devices. Public safety agencies typically share devices among first responders. Therefore, the current user of a device should not be able to access the credentials or information of previous users. The user level security is primarily achieved through various mechanisms.

16.4.1 Device Access Level Security

1. *Using Strong Passwords and Biometrics:* Set a complex password, PIN, or pattern lock, and enable biometric security features like fingerprint or facial recognition. FN/AT&T has developed Identity, Credential, and Access Management (ICAM) to manage user credentials.
2. *Using Two-Factor Authentication:* Enable two-factor authentication on accounts wherever possible to add an extra layer of security.
3. *Device Security Management:* NPSBN has multiuser security management allowing users to share devices without compromising security. FN users are authenticated during checkout of devices and the appropriate configurations distributed to the device. Static configurations can be used for all devices, applied to groups of devices, or configurations can be dynamic based on specific public safety agency policies.
4. *Keeping Software Up to Date:* FN/AT&T regularly updates device operating systems and apps to patch any security vulnerabilities.
5. *Installing Security Apps:* Consider using antivirus and security applications that can help detect and mitigate threats.
6. *Enabling Remote Tracking and Wiping:* Activate features that allow you to track your phone's location and remotely erase data if it is lost or stolen.
7. *Being Cautious with Apps:* Download apps only from FN/AT&T and review permissions to ensure they are not accessing more data than necessary.

16.4.2 Network and Application Access Level Security

Application access governs who can access and interact with applications and associated data. Application access security measures ensure that only authorized users get access to the application and perform certain actions, varying from viewing data to modifying it, based on their roles and permissions. This is crucial for protecting sensitive information and maintaining the integrity of the application. These mechanisms work together to protect user data and prevent unauthorized access to the network, thus ensuring a secure communication environment. Devices which are not FN approved will not be allowed to access NPSBN and FN applications.

1. User Authentication: This involves verifying the identity of users trying to connect to the network, typically utilizing techniques like SIM-based authentication, where the SIM card provides unique identifiers and

keys for authentication. All FN-enabled devices use FN SIM (PLMN 3130100) to access the NPSBN and the desired FN applications.

2. Application Access Authentications: FN applications require user credentials to access, view, and retrieve the application data. NPSBN has specific log-in credentials for FN applications using ICAM.
3. Radio Access Network (RAN) Attachment: This process requires security to access and attach to the FN RAN.

16.4.3 Network Level Security

Network level security involves protecting the network from intruders, whether they are targeted attackers or opportunistic malware. It involves implementing measures to safeguard the integrity, confidentiality, and availability of the network and its data from unauthorized access, misuse, or destruction. Key aspects of network security include implementing firewalls and intrusion detection systems, securing endpoints, using encryption for data transmission, applying strong authentication measures, conducting regular security assessments, and educating users about potential threats like phishing.

NPSBN network level security involves transport level security – air interface between user equipment (UE) and RAN, front backhaul, Cell Site Backhaul (CSBH), Corporate Backbone (CBB), transport to 4G/5G core, IP Multimedia Systems (IMS), and transport connecting external communications networks, e.g., Public Safety Enterprises (PSE) networks which include law enforcement, fire protection, emergency medical services, disaster response, etc.

16.4.4 Air Media Transport Security Measures

Air media transport security measures between the UE and RAN involve encryption and integrity protection. During the establishment of a connection, the UE and the eNB/gNB negotiate encryption keys using the keys derived from the subscriber's credentials stored in the core network. This process ensures that user data and signaling messages are encrypted to prevent eavesdropping, while integrity protection ensures that data cannot be tampered with during transmission.

NPSBN uses 128 EPS Encryption Algorithms1/2 (128-EEA1/2) and 128 EPS Integrity Algorithms ½ (128-EIA1/2) for mobile device authentication and encryption, respectively. 128-EEA1/2 and 128-EIA1/2 are specific security standards. These standards outline procedures and protocols for secure communication among trading partners, emphasizing the protection of data integrity, confidentiality, and authenticity during transmission. These standards are highly followed to handle sensitive information and demand compliance with regulatory requirements. In addition, access control should be in place to prevent unauthorized access to RAN.

16.4.5 Front Backhaul Transport Security Measures

Front backhaul spans between eNB/gNB and Smart Integrated Access Device (SIAD)/Network Termination Equipment (NTE) over fiber/cable lines. An NTE connects to a Mobile Station Node (MSN) via CSBH. SIAD/NTE includes the hardware and software necessary to manage network connections and ensure data transmission. Aggregation Device (SIAD) security features include:

- Access Control List (ACL): ACL features filters and does not allow IPs that are not in the list.
- Malicious Packet Protection: This security measure is designed to detect, prevent, and respond to harmful data packets and block the packet going forward. This includes the use of firewalls, intrusion detection systems (IDS), and intrusion prevention systems (IPS) that analyze network traffic for suspicious behavior, blocking or mitigating threats before they can compromise data integrity or availability.
- Open Shortest Path First (OSPF) Authentications: OSPF allows authenticating OSPF packets to ensure that packets originate from approved nodes. OSPF authentication method is used to secure OSPF routing updates by ensuring that they come from legitimate sources. This authentication helps prevent unauthorized devices from participating in the OSPF routing process and ensures the integrity of the routing information exchanged within the OSPF network.
- Denial of Service (DoS) and DDoS Attack Protection: In DoS attack, a perpetrator uses a single internet connection to either exploit a software vulnerability or flood a target with fake requests – usually in an attempt to exhaust server resources (e.g., RAM and CPU). DDoS attacks are launched from multiple connected devices that are distributed across the internet. These multiperson, multidevice barrages are generally harder to deflect, mostly due to the sheer volume of devices involved. Unlike the

DoS attacks, DDoS assaults tend to target the network infrastructure in an attempt to saturate it with huge volumes of traffic.

- Internet Control Message Protocol (ICMP) Rate Limiter: ICMP rate limiter in NPSBN sets the limits on the number of requests a user or a service can make within a specific timeframe, effectively managing traffic and improving overall service reliability. This helps prevent network congestion and ensures efficient use of resources by limiting excessive unsolicited service requests and assists to mitigate abuse, prevent server overload, and ensure fair resource allocation among users.

16.4.6 CSBH Transport Security Measures

NPSBN's CSBH is based on Metro Ethernet Virtual Local Area Network (VLAN) architecture and spans from the aggregation node (SAID/NTE) to the Mobile Telephone Service Office (MTSO) connecting MSN node. In addition, FN traffic generating from FN/AT&T rural partners are also sent to MSN for onward transmission to FN core. The NPSBN critical security features in MSN include ACL restricting IP traffics and allowing only S1 U, S1 C, and operations, administration, and maintenance (OAM). MSN has all other security features in place similar to SIAD-enabled MSN to control and manage the security of the traffic coming from rural partners.

16.4.7 NPSBN MPLS Backbone Transport Security Measures

NPSBN's IP Multiprotocol Label Switching (MPLS) backhaul transport network (CBB) has several security features, including traffic segregation through label-based routing, which minimizes exposure to malicious traffic. It supports Virtual Private Networks (VPNs), allowing for private, isolated communication channels within a shared infrastructure. In addition, FN traffics (S1 U and S1 C) are encrypted to protect data integrity and confidentiality during transmission over CBB. The network's ability to control Quality of Service (QoS) enables sensitive data to be prioritized and transmitted more securely.

NPSBN utilizes Terminal Access Controller Access-Control System (TACACS) protocol used for Authentication, Authorization, and Accounting (AAA) remote networking access services. It enables centralized control over user access to network resources by allowing devices like routers and switches to communicate authentication requests to a TACACS server.

TACACS provides features such as accounting and authorization in addition to authentication, with a focus on security measures through encryption. Global TACACS (GTAC) plays a critical role in controlling access to network elements and the principle of least privilege. GTAC enforces user password complexity and periodic password expiration. FN/AT&T GTAC team manages the GTAC system. GTAC security measure includes:

- *Data Encryption:* LTE employs encryption algorithms, e.g., Advanced Encryption Standard (AES) to secure the user data transmitted over the air, ensuring that even if the data is intercepted, it remains unintelligible without the decryption keys.
- *Integrity Protection:* This ensures that the data sent from users is not altered during transmission. LTE uses integrity protection to safeguard against vulnerabilities and tampering in the control signals and user data.
- *Multiprotocol Virtual Private Network (MVPN):* This technology enables secure communication across a shared public infrastructure while allowing multiple endpoint connections through a single network. It supports multiple VPN types and can efficiently route data while maintaining privacy and security.

16.4.8 Information and Data Security

NPSBN information and data security areas include RAN and core security measures.

16.4.8.1 RAN Security Measures

NPSBN's RAN security measures include the use of authentication mechanisms, encryption, and integrity protection. These measures aim to protect user data and ensure secure communication between the UE and RAN over air and the NPSBN core over transport media.

1. Authentication: Users are authenticated using identity verification techniques, typically involving the Subscriber Identity Module (SIM) and the use of the mutual authentication process between the UE and the network.
2. Integrity protection: This technique ensures that the data is not altered during transmission, is protected from tampering, and that signaling

continues between the same elements as verified by authentication. The integrity check is done within the RAN.
3. Encryption: Data transmitted over the air is encrypted using algorithms such as AES to prevent eavesdropping.
4. Radio interface ciphering function: A cipher is a method used to encrypt or encode information, transforming plaintext into a format that is not easily readable (cipher text). Deciphering is the process of converting the cipher text back into its original plaintext form. These processes rely on algorithms and keys, ensuring that only authorized users can access the original message.

16.4.8.2 Core and Interface Security Measures

NPSBN core is the center of all data processing and information sharing between NPSBN users and applications. NPSBN's Mobility Management Entity's (MME) main functions include controlling the signaling for mobile devices, managing UE registration, and handling mobility management procedures. Specifically, the MME is responsible for bearer path management, session management, and authentication of users. It plays a critical role in ensuring efficient communication between the UEs and the core network and have critical built-in security features including IP network separation, IP filtering, DoS/DDoS attack prevention measures, overload protections, and malicious packet protections listening only to a specific port (port 36412).

NPSBN's System Architecture Evolution Gateway (SAE-GW) handles all FN user data packet transmission between user devices and various data sources and limits the use, dissemination, and access of restricted data to specific agencies, individuals, and situations listening to port 2152. NPSBN security measure complies to the Third Generation Partnership Project (3GPP) Security Gateway (SGW) Internet Protocol security (IPsec) and certificate management specifications to provide access control through authentication and traffic confidentiality and integrity through encryption.

HSS, PCRF, PCEF, and other nodes store user profile, QoS parameters, and other Customer Proprietary Network Information (CPNI), Personally, Identifiable Information (PII), and ensitive Personal Information (SPI) data. Core nodes are secured and monitored for security breach and logs are maintained as part of the security compliance.

16.4.8.3 IMS Security Measures

In Session Initiated Protocol (SIP) sessions, IPsec is used for VoLTE-SIP messages. SIP call signals are kept confidential to protect user identities, network topology, and call routing information which could be used to gain unauthorized access to services. Access control is used to prevent unauthorized access to IMS elements. Session initiation protocol (RFC 3261)supports the SIPS URI scheme in accordance with updated RFC 5630.

Interface and external interface security measures include:

1. FW protection at external interfaces
2. IPS/IDS at external interfaces
3. Malware at external interfaces

Secure Real-time Transport Protocol (RTP) (or SRTP) is the primary mechanism used for media encryption. It is an RTP profile that provides:

- Media confidentiality through fast AES encryption (a stream cipher)
- Media packet integrity using message authentication codes
- Media packet replay protection (protects against DoS packet flooding attacks)
- Key management framework (which provides forward security)

SIP uses DNS procedures to allow a client to resolve a SIP Uniform Resource Identifier (URI) into the IP address, port, and transport protocol of the next hop to contact. It also uses DNS to allow a server to send a response to a backup client if the primary client has failed (75).

16.4.8.4 External Interface Security Measure

NPSBN's external interface protection measures are in place to safeguard all external interfaces with appropriate security measures such as firewalls, protection from common internet attack vectors (e.g., DOS, DDoS, spoofing, malware, botnets, port scanning), intrusion prevention and detection, security gateways, security logging, and content inspection/filtering. External interfaces include:

- Service Gateway Interface (SGi) interfaces
- Roaming interfaces such as S8 and S6a

- Public Switched Telephone Network (PSTN) and Voice over IP (VoIP) peering for voice and messaging traffic
- Public Safety Enterprise Networks (PSEN) interfaces
- Applications ecosystem interfaces with content providers, application developers, and service providers offering services via the applications ecosystem

16.4.9 NPSBN Cybersecurity

The NPSBN cybersecurity measures include a dedicated cybersecurity program that considers all source threats; constructs a dynamic threat profile; generates a cybersecurity architecture; builds proactive forensics; and establishes incident response capabilities that ensure the ability to operate and deliver crucial services as needed during a national, state, or local incident.

The NPSBN cybersecurity measures ensure that the NPSBN is protected from cyberattack, yet public safety users can readily access the network. All critical operational equipment and functions, which could affect the secure and effective operations of the NPSBN, are located within the sole jurisdiction of the United States. To that end, the solution considers the following areas: usability – The network should be usable by PSEs; security controls, policies, and procedures should provide protection without impacting operability or interoperability.

16.4.10 Cybersecurity Framework

Cybersecurity Framework (CSF) is a set of voluntary guidelines designed to help organizations assess and improve their ability to prevent, detect, and respond to cybersecurity risks. The framework, developed by the U.S. National Institute of Standards and Technology (NIST), was first published in 2014 for critical infrastructure sectors and has since been broadly adopted across various industries (see Figure 16.2).

NIST cyber security framework is designed to be flexible and adaptable, making it useful for organizations of various sizes and industries to enhance their cybersecurity measures. The framework consists of five core functions:

- *Identify:* Understand your organization's cybersecurity risks, assets, data, and capabilities
- *Protect:* Develop safeguards to ensure critical infrastructure services can deliver their function

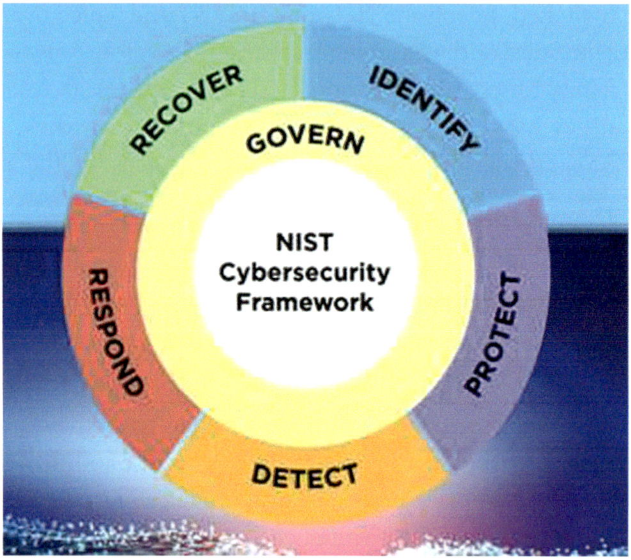

FIGURE 16.2
NIST Cyber Security Framework.

- *Detect:* Implement activities to identify the occurrence of a cybersecurity event
- *Respond:* Act immediately after detecting a cybersecurity incident
- *Recover:* Develop and implement plans for resilience and restore any capabilities or services impaired during a cybersecurity incident

Security features in Security Information and Event Management (SIEM) systems typically include:

1. *Log Management:* Centralized collection, storage, and analysis of logs from various sources.
2. *Real-Time Monitoring:* Continuous analysis of security events to detect anomalies or potential threats.
3. *Incident Response:* Automated or manual workflows for responding to detected security incidents.
4. *Threat Intelligence Integration:* Incorporation of external threat data to enhance event correlation and detection capabilities.
5. *User Behavior Analytics:* Monitoring user actions to identify unusual behavior that may indicate a breach.
6. *Compliance Reporting:* Tools to assist organizations in meeting regulatory compliance requirements by generating necessary reports.

17

FN Operations and Emergency Management

The Nationwide Public Safety Broadband Network (NPSBN) operations and emergency management programs are initiatives designed to provide reliable and secure wireless communication for first responders and emergency management personnel across the United States. These programs focus on enhancing collaboration, improving situational awareness, and ensuring effective communication during emergencies. They utilize advanced technologies and networks to facilitate real-time data sharing and coordination among various agencies.

17.1 NPSBN OPERATIONS 24×7×365

The NPSBN 24×7×365 operation plans focus on ensuring continuous, reliable communication services for public safety agencies. These plans typically include system redundancies, around-the-clock monitoring, rapid troubleshooting, and support teams available to address any issues that may arise, thereby maintaining the network's integrity and performance.

NPSBN is based on and fully adhere to the Fault Management, Configuration Management, Accounting Management, Performance Management, and Security Management (FCAPS) specifications standardized by the International Telecommunication Union (ITU), which defines the framework for managing telecommunications networks. It outlines effective monitoring, control, and management of services and network elements to ensure quality and reliability. Fault and performance-related issues may be at the user level or network level.

DOI: 10.1201/9781003557739-19

Fault Management and Performance Management in NPSBN operations maintain the network health and performance. NPSBN's network monitoring systems collected data directly from the nodes that provide the service and correlate it to the network inventory in order to accurately show where faults and performance issues exist. To ensure proper network security while permitting authorized personnel to perform troubleshooting, access is managed at the employee level within the NPSBN. FirstNet's American Telephone and Telegraph Company (FN/AT&T) uses auto-ticketing systems to send repair tickets to technicians.

Performance Management systems provide the infrastructure for real-time alerting and monitoring of NPSBN's performance to the operation center. The Ops personnel proactively detect any network problem and repair in real time by collecting measurements from network elements and processing them against predefined thresholds for exception alerting. FN performance reports on specific services and customer performance reports are based on the data collected from network elements.

Trouble Ticket Management

NPSBN trouble ticket management systems are designed to oversee and track technical issues either reported by users of the NPSBN or issues detected in the network by Ops. This system aids in efficiently managing incidents, ensuring that problems are resolved in a timely manner, and helps maintain the network's operational integrity for public safety communications.

FN trouble tickets are managed and resolved by AT&T One Ticketing System Trouble Management (AOTS-TM) which uses a single end-to-end ticket ecosystem that enables enhanced ticket correlation, business intelligence, and cross-team interoperability.

FN/AT&T has a dedicated phone line for first responders to reach FN Ops personnel for any technical issues.

FN Emergency Management Coordination and Event Management

FN was created to provide reliable, high-speed communications during emergencies and disasters, ensuring that first responders have the tools they need to perform their duties effectively. Types of emergencies and disasters include natural disasters (such as earthquakes, floods, wildfire, and hurricanes), man-made disasters (e.g., chemical spills, terrorism, and nuclear accidents), public health emergencies (including pandemics and outbreaks), and technological incidents (power outages and transportation accidents). Each category can vary significantly in impacts and response requirements and must be managed promptly and efficiently with a coordinated efforts among federal agencies, local government agencies, and FN personnel.

17.2 FEDERAL EMERGENCY MANAGEMENT AGENCIES

Federal Emergency Management Agencies (FEMA) (76) is a U.S. federal government agency under the U.S. Department of Homeland Security (DHS) and is responsible for emergency management and support individuals and communities hit with the disasters. FEMA is called upon to lead the coordination of federal efforts to stabilize communities after a disaster, and to make individuals and communities to become resilient than before.

FEMA works closely with FN and states/local government, tribal nations, and territories as they make requests for federal support. The National Disaster Recovery Framework of FEMA enables a whole community approach and allows agencies to coordinate with one another to provide communities unified federal assistance as quickly and effectively as possible.

17.3 FN RESPONSE OPERATION GROUP

The FN Response Operation Group (ROG) is a dedicated team created to provide support for emergency response and public safety communications, primarily utilizing the FN network resources. This group plays a crucial role in assisting first responders during disasters and major incidents by ensuring they have reliable and secure communication capacities for planned events (e.g., Super Bowl games, New Year celebrations in Time Square and Thanksgiving parade in New York City) and emergencies. The ROG develops programs and policies, internal and external relationships, and long-range planning and strategy to meet FN missions.

The group is led by the director of the ROG for the FN program at AT&T. ROG has dedicated personnel in each FEMA region to coordinate on behalf of FN and provide FN resources during event management including FN deployables based on the request from the public safety agencies for FN coverage.

17.4 NPSBN PROGRAM OVERSIGHT

The Oversight Board (2) was established to supervise and guide FirstNet, which is tasked with developing, building, and operating the NPSBN. The board ensures that FN meets the needs of public safety agencies and stakeholders, and it typically comprises representatives from various sectors, including public safety organizations, telecommunications industry, and federal agencies.

In addition, within FN, a team engages with AT&T teams for technical issues, provides program guidance and oversight to the FN, and manages program coverage and operations with the FN program team in AT&T.

Section III

NPSBN Future Innovations, Opportunities, and Challenges in Public Safety

18

NPSBN Operational Innovations and Interoperability

Nationwide Public Safety Broadband Network (NPSBN) faces several future challenges. These challenges require continuous investment, innovation, and cooperation with public safety leaders to ensure the network remains robust and reliable for first responders for the foreseeable future. A few imminent opportunities and challenges are briefly described in this chapter.

18.1 OPERATIONAL INNOVATIONS

Operational innovations refer to new methods or improvements in systems, processes, or practices within an organization that enhance efficiency, reduce costs, or increase service quality. These innovations lead to significant changes in how an enterprise operates, allowing for better performance and competitive advantage.

FirstNet is continuously working and coordinating with National Institute of Standards & Technologies (NIST), Federal Emergency Management Agency (FEMA), device and network vendors, and public safety agencies to enhance the operational capabilities of first responders.

18.1.1 Developing Additional Applications and Integrating with NPSBN Networks

FirstNet and American Telephone and Telegraph Company (AT&T) has useful applications available on the FirstNet (FN) application catalog site. Third parties develop these applications and FirstNet certifies them. First responders download any applications on their devices and use them as required. FN applications encompass mobile device management, Mission

DOI: 10.1201/9781003557739-21

Critical Push to Talk (MCPT), and navigation-related areas. These applications are available for different operating systems.

With the advent of 5G/6G network technology, first responders require applications to meet the needs such as Augmented Reality (AR), Artificial Intelligence (AI), Low Earth Orbit (LEO)–supported satellite applications, Internet of Things (IoT), self-driving vehicles, etc.

18.1.2 Enhancing FN Devices for 6G and Other Relevant Technologies

Mobility communication innovations are constantly improving and now moving to the sixth generation, which is set to succeed 5G. It aims to deliver higher data rates, lower latency, and increased connectivity, potentially supporting advanced applications such as immersive augmented and virtual reality, holographic communications, and IoT. While 6G is still in the research and development phase, it is anticipated to roll out around 2030, building on and expanding the capabilities of 5G networks.

FirstNet and AT&T developed and deployed NPSBN and have large options of mobile devices for 4G/5G available to first responders. The next generation of devices are in the standardization and development cycles. Apple, Samsung, Ericsson, Nokia, Intel, and many other companies are actively participating in 6G standardization and development, aiming for capabilities in 6G phones. FN/AT&T should proactively engage with mobile device manufactures at the early development lifecycle to include FirstNet features and services in the 6G-capable devices including 6G-connected cars, VR, AI-based network intelligence for fault and performance alarms and monitoring, duel smartphones that can connect both terrestrial cell and LEO satellites that do not need deployables.

18.1.3 Blockchain for Network Security and Authentication

Blockchain technology for communication network design (77) is used to enhance security, transparency, and efficiency in data transmission. It allows decentralized and encrypted communication by storing information in distributed nodes, making it difficult for unauthorized parties to alter data. This improves trust among users and reduces reliance on central authorities, thereby facilitating peer-to-peer interactions. In general, the communication networks blockchain covers various communication networks such as cellular networks, Wi-Fi, cloud computing, IoT, software-defined network, and smart networks.

Blockchains are distributed digital ledgers of cryptographically signed transactions that are grouped into blocks. Each block is cryptographically linked to the previous one (making it tamper evident) after validation and undergoes a consensus decision. As new blocks are added, older blocks become more difficult to modify (creating tamper resistance). New blocks are replicated across copies of the ledger within the network, and any conflicts are resolved automatically using established rules.

For NPSBN, "Permissioned Blockchain" network setup, run and govern by FN/AT&T is highly desirable rather a "Permissionless Blockchain" network which is often governed by blockchain network users, publishing nodes, and software developers. Each group has a level of control that affects the direction of the blockchain network's advancement.

Blockchain technologies (78) are touted as extremely secure due to their tamper-resistant design – once a transaction is committed to the blockchain, it generally cannot be changed. To start with, FN/AT&T should actively invest in blockchain technology and include the design into NPSBN especially in dealing with cybersecurity to enhance security in user authentications, authorizations, and access to data.

Blockchain should include NPSBN user devices and subsystems in architecting the blockchain-based network and applications:

1. Users: Mobile devices and IoT devices that connect to the LTE network.
2. Base Stations: eNB/gNB that interface with users, collecting data, and sending it to the core network.
3. Core Network: Mobility Management Entity (MME), Serving Gateway (SGW), and Packet Data Network Gateway (PGW), which manage mobility and data services.
4. Blockchain Network: A decentralized ledger technology layer that can be used for secure transactions, identity management, and trustless interactions between different entities in the LTE ecosystem.
5. Smart Contracts: Automated contracts that execute transactions based on defined criteria, facilitating operations like data sharing and billing.
6. APIs and Interfaces: For interaction between the LTE network and blockchain services, as well as external applications.

18.2 AI FOR NETWORK/SERVICE ENHANCEMENT AND MONITORING

AI is enticing communities in many fields, including the technology. AI is being significantly pursued by business sectors such as healthcare for diagnostic and treatment assistance, finance for fraud detection and investment analysis, retail for inventory management, transportation for autonomous vehicles, and communications for customer service, and virtual assistants. Its versatility allows for automation and enhanced decision making across various fields. AI is the simulation of human intelligence in machines that are programmed to think like humans and mimic their actions. AI systems can perform tasks that require human-like abilities, such as understanding natural language, recognizing patterns, solving problems, and learning from experience.

AI can immensely enhance network and service monitoring by using machine learning algorithms to analyze vast amounts of data from network devices, applications, and users in real time. This analysis allows for anomaly detection, performance optimization, predictive maintenance, and automation of responses to network issues. AI can identify patterns in network traffic that indicate problems and alert operators or automatically take corrective actions, improving overall service reliability and efficiency.

FN/AT&T should initiate systemic approach to incorporate AI into different segments of NPSBN, especially network inventory and network resource assignment based on user profiles and current assignment scenarios, network monitoring, alarming, and cyber security. AI can significantly enhance the NPSBN by improving data analysis, optimizing resource allocation, and enhancing communication among emergency responders. AI can analyze large volumes of data rapidly to identify trends, predict incidents, and optimize routing for emergency responses. It can also streamline communication through automated systems, ensuring that first responders receive timely updates and information during critical situations.

18.3 ENHANCING 9-1-1 SERVICES INTEROPERABILITY AND 4G/5G TECHNOLOGY INTEROPERABILITY

In the United States, 9-1-1 is the emergency telephone number that connects callers to local Public Safety Answering Points (PSAPs). Each PSAP is responsible for receiving emergency calls and dispatching the appropriate emergency services. There are approximately 5,400 PSAPs across the country having various levels of features, facilities, and infrastructures.

According to the national 9-1-1 progress report 2020 data, an estimated 240 million emergency calls are made regularly. About 80% of the calls to 9-1-1 are made from mobile phones, and the rest are still using PSTN or Voice over Internet Protocol (VoIP). In basic 9-1-1 calls, three-digit number (9-1-1) is dialed and a call taker/dispatcher in the local PSAP call center answers the call. The emergency and its location are communicated by voice (or TTY) between the caller and the call taker. In a PSAP that has enhanced 9-1-1 (e9-1-1), the call is selectively routed to the proper PSAP for the caller's location, and the PSAP has equipment and database and also information that displays the caller's phone number and address to the call taker. Currently, 93% of counties with 9-1-1 coverage have enhanced 9-1-1 for callers.

The term "enhanced 9-1-1" is not synonymous with wireless 9-1-1 but may use VoIP telecommunications services. The National Emergency Network Association (NENA) is taking the lead to develop full Next Generation 9-1-1 (NG9-1-1) capability for VoIP-based services. NENA is not a government organization; it is a nonprofit organization focused on promoting the implementation of effective emergency communication systems and developing 9-1-1 standards for handling 9-1-1 calls more efficiently.

With the recent works in wireless emergency calls, call takers receive both the caller's wireless phone number and their location information. The call is routed to a PSAP either based on cell site/sector information or on caller location information. 9-1-1 services in most states are increasingly being integrated with NG9-1-1 systems, which allow for enhanced capabilities including text and data transmission.

American National Standards Institute (ANSI), national Alliance for Telecommunications Industry Solutions (ATIS), NENA, NIST and many other standards bodies are actively working to develop NG9-1-1 standards and national collaboration to utilize and monitor progress on the options listed below to address standards, technological barriers, and issues identified in the National Plan for Migrating to IP-enabled 9-1-1 systems to ensure:

- Complete and accept IP-enabled 9-1-1 open standards and understand future technology trends to encourage system interoperability and emergency data sharing
- Establish routing and prioritization protocols and business rules

- Determine the responsible entity and mechanisms for location acquisition and determination
- Establish system access and security controls to protect and manage access to the IP-enabled 9-1-1 system
- Develop a certification and authentication process to ensure service providers and 9-1-1 authorities meet security and system access requirements.

FN/AT&T should actively participate in the national 9-1-1 standards development process and shape the NG 9-1-1 standards based on the achievement of NPSBN program and its successful implementations in the United States and the territories. FN/AT&T should highlight the goal of NPSBN in saving lives and properties, and PSAP enhancements to integrate NPSBN features for location acquisition and FN QPP-based routing would be part of the overall PSAP enhancement integration.

18.4 EMERGENCY MEDICAL SERVICE NETWORK SPECIFIC CHALLENGE

Emergency Medical Service (EMS) is vital in saving lives during an emergency. The goals of an EMS network are primarily to provide timely and effective medical assistance during emergencies, ensure efficient transportation of patients to appropriate medical facilities, and improve overall patient health outcomes through coordinated healthcare services. EMS networks also aim to enhance public safety, increase community awareness about health services, and facilitate disaster response and preparedness.

EMS (79) encompasses the initial stages of the emergency care continuum. It includes emergency calls to 9-1-1, transport of patients by ambulance and air medical service. The speed and quality of emergency medical services are critical factors in saving a patient's life. For patients who cannot breathe, are in hemorrhagic shock, or are in cardiac arrest, the decisions made and actions taken by EMS personnel may determine the outcome as much as the subsequent hospital-based care and may mean the difference between life and death.

EMS networks today face several challenges in technology integration and interoperability, data sharing, and communications besides funding and staffing shortages.

Ensuring seamless communication and incident-related data sharing among PSAP, police, fire, and EMS including off-site healthcare providers are often hindered by incompatible systems. Sharing the incident location, nature, and scope of the emergencies, video/text and information relating to incident command and protocols, etc., among incident management team including EMS will highly enhance the emergency management and provide timely medical relief to people impacted by emergencies. NPSBN should provide a base platform and framework for the overall integration and interoperability of the EMS communications and incident data sharing among PSAP, police, fire, and EMS.

18.5 INCORPORATING LEO TO ENHANCE NETWORK PERFORMANCE

LEO is the orbit around Earth with an altitude typically ranging from about 99–1,240 miles (160–2,000 kilometers) above the Earth's surface. Satellites in LEO can complete an orbit in approximately 90–120 minutes, making it a preferred orbit region for communications, Earth observation, and space missions. In Geosynchronous Earth Orbit (GEO), satellites orbiting at about 22,366 miles (35,787 km) have a period precisely equal to one day. Satellites in this orbit observe the Earth as if the Earth is not rotating. Because of this property, satellites in GEO are constantly in the field of view for approximately one-third of the planet's surface.

FirstNet deployables uses Inmarsat satellites which typically operate in GEO, approximately 22,236 miles (35,786 kilometers) above the Earth's equator. This allows Inmarsat satellites to maintain a fixed position relative to Earth's surface, providing consistent coverage to specific geographic regions. Although Inmarsat deployed in GEO has certain specific advantages of constant exposure in the field of view for approximately one-third of the planet's surface, the data packet latency is significantly higher – approximately about 750 ms on average. Starlink, which is deployed in LEO, typically offers lower data transmission latency compared to Inmarsat. Starlink, utilizing a constellation of LEO satellites, can achieve lower latencies around 30–45 ms compared to Inmarsat. The data packet latency of Inmarsat services typically ranges from 700 ms to 750 ms, depending on various factors such as the specific satellite network, weather conditions, the end-user device, and the higher altitude.

Starlink's growing network of LEO satellites continued to deliver relatively low latencies, which are important for apps such as videoconferencing, when compared to GEO systems. As per Ookla, Starlink's median latency in Q1 2021 was 45 ms. FN/AT&T has already started investigation and testing of various deployables on Starlink. Hopefully, all other features of FN deployables would be compatible over LEO along with unconstrained attachment of the existing and future FN-approved devices.

18.6 MARITIME RESPONSE AND INTEROPERABILITY

Maritime has not yet been in the active phase of discussion to include into the NPSBN program. The author has personal working experience and some preliminary discussions with the first responders while working with them in Guam, American Samoa, U.S. Virgin Islands, and Puerto Rico. It has been observed that more often than not, these islands get sever hurricanes damaging the NPSBN RAN, power lines, flooding the roads, etc., leaving the islands completely isolated for days and weeks. Limited resources of deployables are not sufficient and cannot provide NPSBN connectivity to all first responders in managing disasters.

Maritime solutions can provide much-needed help in situations where all or majority of terrestrial cells are damaged. A fleet of ships equipped with NPSBN RAN attaching to satellite networks can be anchored at suitable locations closer to the islands and would provide communication service support to a larger group of the first responders to enable them to perform the disaster recovery. A feasibility study should be performed to identify the necessary enhancements to deploy 4G/5G RAN on ships, satellite interface requirements and connectivity to satellites and the 4G/5G cores, network security considerations, FN device requirements/constraints, if any, and testing and validation.

18.7 INCORPORATING FNA/NPSBN REQUIREMENTS INTO NATIONAL AND INTERNATIONAL STANDARDS

NPSBN has passed the initial critical phase of development and deployment. It is now effectively serving the public safety communities and is ready to lead the public safety network standardization efforts for emergency services across the nation. This standardization enhances interoperability among various public safety agencies, ensuring they can efficiently communicate and coordinate during emergencies. The NPSBN's architecture supports robust, high-speed data transmission, which is essential for real-time information sharing among first responders, thus improving the overall public safety response capabilities. It is of utmost importance that NPSBN-related technical knowhow and the methods and procedures become the standard operating procedure for entire public safety communities.

18.8 NPSBN INTEROPERABILITY AND CHALLENGES

NPSBN interoperability allows different emergency service organizations, such as police, fire, and emergency medical services, utilities companies, etc., to communicate seamlessly with one another using modern broadband technologies and integrate with their older device and technology to function cohesively with NPSBN. This interoperability is crucial for effective emergency response and coordination across various jurisdictions and agencies.

18.8.1 MCPTT-Land Mobile Radio Interoperability

Land Mobile Radio (LMR) system is a type of two-way radio communication system designed for mobile and portable users, typically used by public safety organizations for mission-critical voice communication – Push-to-Talk two-way communication between two radio transceivers. LMR operates on various frequencies, allowing users to communicate over short-to-medium distances, often using repeaters to extend range. Currently, a significant number of public safety agencies, especially police and fire, have LMR system that need to be integrated with the NPSBN MCPT services to provide full advantage of NPSBN's MCPT services to these agencies.

FN/AT&T plans to support Mission Critical Push to Talk (MC-PTT) to LMR interoperability as part of MC-PTT product feature as an independent feature for MC-PTT subscribers. This should enable public safety agencies to make use of their existing LMR infrastructure and enhance their capabilities to use NPSBN and LMR both as needed. The MC-PTT LMR interoperability feature will be supported through Radio over Internet Protocol (RoIP), offering a straightforward and dependable method for integrating LMR and MC-PTT networks.

18.8.2 NPSBN Interoperability with Other Service Providers

Currently, the NPSBN is created, implemented, and managed by AT&T through a public-private contract selection process. However, there are other mobile service providers serving public safety agencies on their 4G/5G networks with no interoperability with NPSBN. Public safety agencies may have long-term contracts or other reasons to stay with their current mobile network providers rather than switch to FN/AT&T for NPSBN services designed for first responders. This situation would possibly hamper the ability of these public safety agencies to take full advantage of the NPSBN service features, QPP, and other emergency management features thus depriving the public in general served by such agencies of the ability to effectively manage emergency situations and save lives and properties.

NPSBN interoperability to all mobile service providers would certainly be helpful but it deserves a broader view and full extent of the impact and longevity of the NPSBN, funding issues, customer ownership, and sharing of the revenue among FN/AT&T and other mobile operators interoperating with NPSBN to provide NPSBN services to public safety agencies and first responders. This is an open issue and needs a candid, careful, and diligent approach.

Glossary

Acronym	Definition
3G	Third Generation
3GPP	3rd Generation Partnership Project
3PT	Three Pacific Territories
AAA	Authentication, Authorization, Accounting
AC	Access Class
ACB	Access Class Barring
ACL	Access Control List
AF	Assured Forwarding
AH	Authentication Header
AI	AI Artificial Intelligence
AKA	Authentication and Key Agreement
ALI	Automatic Location Information
AM	Acknowledged Mode
AMBR	Aggregated Maximum Bit Rate
ANR	Automatic Neighbor Relations
ANT	Automated Network Testing
AOTS	AT&T One Ticketing System
APCO	Additional Protocol Configuration Options
API	Application Programming Interfaces
APN	Access Point Name
APN-DNS	Access Point Name – Domain Names System
AR	Augmented Reality
ARP	Allocation and Retention Priority
ASN	Autonomous System Number
ATCA	Advanced Telecommunications Computing Architecture
AVPN	AT&T Virtual Private Network
BAU	Business as Usual
BBM	Base Band Module
BBU	Base Band Unit
BFD	Bidirectional Forwarding Detection
BGP	Border Gateway Protocol
BSS	Business Support Systems
BYOD	Bring Your Own Device

c-PCRF	Common Policy Charging Resource Function
CA	Carrier Aggregation
CALEA	Communications Assistance for Law Enforcement Act
CAP	Common Alerting Protocol
CAS	Common Access Services
CBB	Common Backbone Network
CDR	Call Detail Records
CF	Control Functions
CE	Customer Edge
CMS	Content Management Server
CN	Core Network
COAM	Customer Owned and Maintained
COLT	Cells on Light Trucks
CONUS	Congruent United States
CoS	Class of Service
COW	Cells on Wheels
CPNI	Customer Proprietary Network Information
CRD	Compact Rapid Deployable
CTS	Converged Telephony Server
CSBH	Cell Site Backhaul
CSCF	Call Session Control Function
CSFB	Circuit Switch Fall Back
CUPS	Control User Plane Separation
DAS	Distributed Antenna Systems
DDN	Downlink Data Notification
DEA	Diameter Edge Agent
DHCP	Dynamic Host Configuration Protocol
DL	Downlink, Downloadable
DNS	Domain Name Service
DPCC	Dynamic Policy Charging and Control
DRA	Diameter Routing Agent
DRS	Discovery Reference Signal
DSCP	Differentiated Services Code Points
E911	Enhanced 911
EAP	Extensible Authentication Protocol
EEA	EPS Encryption Algorithms
EIA	Electronic Industries Alliance
ELEAT	Electronic Law Enforcement Activity Tracker
EMBB	Enhanced Mobile Broadband

EMG	Email Gateway
EMM	Enterprise Mobility Management
EMS	Emergency Medical Services
eNB	Enhanced Node Base Station
eNodeB	Evolved Node B
EAP-AKA	Extensible Authentication Protocol-Authentication and Key Agreement
EOC	Emergency Operations Centers
EPC	Evolved Packet Core
ePDG	Enhanced Packet Data Gateway
EPS	Evolved Packet System
EPU	Extended Primary Users
E-RAB	E UTRAN Radio Access Bearer
ESF2	Emergency Support Function 2
ESInet	Emergency Services IP Network
ESP	Encapsulated Security Protocol
EUTRAN	Evolved UMTS Terrestrial Radio Access Network
EVC	Ethernet Virtual Circuit
EXP	Experimental
FAQ	Frequently Asked Questions
FCAPS	Fault Configuration Accounting Provisioning Security
FCC	Federal Communication Commission
FEMA	Federal Emergency Management Agency
FFA	First Field Application
FIPS	Federal Information Processing Standard
FISMA	Federal Information Security Management Act
FM	Fault Management
FMBT	FirstNet Mobility Tools
FIRSTNET	FirstNet
FIRSTNETRP	FirstNet Rural Partner
FPSS	First Preemption SPID Set
FQDN	Fully Qualified Domain Name
FRR	Fast ReRoute
FTP	File Transfer Protocol
FTTC	Fiber to The Cell
FTTH	Fiber to The Home
FW	Firewall
GBPS	Gigabit Per Second

GBR	Guaranteed Bit Rate
GCF	Global Certification Forum
GCS	Group Communications Services
GCSE	Group Communication System Enablers
GETS	Government Emergency Telecommunications
GMLC	Gateway Mobile Location Centre
GMOP	Global Method of Procedure
GMS	Group Management Server
GNOC	Global Network Operations Center
GPRS	General Packet Radio Service
GRT	Global Response Team
GSM	Global System for Mobile
GSMA	Global System for Mobile Association
GTAC	Global TACACS
GTOC	Global technology operations Center
GTP-C	GPRS Tunneling Protocol Control
GTR	Global Response Team
GUTI	Global Unique Temporary UE Identify
HazMat	Hazardous Materials
HetNet	Heterogeneous Network
HO	Hand Over
HPA	High Priority Access
HPLMN	Home Public Land Mobile Network
HPUE	High Powered User Equipment
HSPA	High Speed Packet Access
HSS	Home Subscriber Server
HTTP-FT	Hypertext Transfer Protocol – File Transfer
HTTPS	Hypertext Transfer Protocol - Secured
IANA	Internet Assigned Numbers Authority
IC	Incident Commander
ICAM	Identity Credential and Access Management
ICIC	Inter-Cell Interference Coordination
ICLB-L	Inter Carrier Load Balancing
ICMP	Internet Control Message Protocol
ICSCF	Interrogating Call Session Control Function
IdM	Identity Management
iDMS	Identity Management Server
IdP	Identity Provider
IDS	Intrusion Detection System

IEEE	Institute of Electrical and Electronics Engineers
IETF	Internet Engineering Task Force
IFO	Inter Frequency Offload
iHSS	Integrated Home Subscriber Server
ILEC	Incumbent Local Exchange Carriers
IMEI	International Mobile Equipment Identity
IMMCI	Idle Mode Mobility Control Info
IMS	IP Multimedia Subsystem
IMS-AS	IP Multimedia Subsystem – Application Server
IMSI	International Mobile Subscriber Identity
IOPS	Isolated Operations for Public Safety
IoT	Internet of Things
IP	Internet Protocol
IP-CAN	Internet Protocol – Connectivity Access Network
IPAG	IP Aggregation Gateway
IPAWS	Integrated Public Alert and Warning System
IPME	IP Messaging
IPS	Intrusion Prevention System
IPSEC	Internet Protocol Security
IPX	Internetwork Packet Exchange
IR94	IMS Profile for Conversational Video Service
IRAT	Inter Radio Access Technology
ISO	International Organization for Standardization
ITIL	Information Technology Infrastructure Library
ITP/IRS	IMEI Trigger Platform / IMEI Reporting Server
ITSC	Information Technology Service Continuity
ITU–T	International Telecommunication Union – Telecommunication Standardization Sector
IXC	Inter Exchange Carrier
KPI	Key Performance Indicators
LAA	Licensed Assisted Access
L2/L3	Layer 2 / Layer 3
L2VPN	Layer 2 Virtual Private Network
LACP	Link Aggregation Protocol
LAG	Link Aggregation Groups
LB	Load Balancing
LBO	Local Break Out
LBS	Location Based Service
LBT	Listen Before Talk

LDAP	Lightweight Directory Access Protocol
LIG	Lawful Intercept GW
LIMD	Lawful Intercept Module
LIPD	Lawful Intercept Packet Data
LMA	Local Mobility Anchor
LMR	Land Mobile Radios
LOA	Levels of Assurance
LRN	LNP Routing Number
LTE	Long Term Evolution
MAC	Media Access
M2M	Machine to Machine
MBPS	Megabit Per Second
MBR	Maximum Bit Rate
MC	Mission Critical
MCPTT	Mission Critical Push to Talk
MCS	Mobility Connected Services
MCS-AS	Mission Critical Service – Application Server
MCV	Mission Critical Voice
MCX	Mission Critical Service
MDM	Mobile Device Management
MEF	Metro Ethernet Forum
MGW	Media Gateway
MIMO	Multiple Input Multiple Output
MIND	Master Integrated Network Directory
MIPE	Mobility Infrastructure Provider Edge
MIS	Managed Internet Service
MME	Mobility Management Entity
MMG	Mobile Messaging Gateway
MMS	Multimedia Messaging Service
MNO	Mobile Network Operator
MO	Mobile Originated
MOB DNS	Mobility Domain Name System
MOCN	Multi Operator Core Network
MOP	Method of Procedure
MPLS	Multiprotocol Label Switching
MPS	Multimedia Priority Services
MRF	Media Resource Function
MRFC	Media Resource Function Controller
MRFP	Media Resource Function Processor

MSA	Mobile Security Alliance
MSB	Master Subscriber Database
MSC	Mobile Switching Center
MSISDN	Mobile Station International Subscriber Directory Number
MSN	Mobile Service Node
MSN	Multiservice Node
MSP	Multi Service proxy
MSS	Mobile Satellite Services
MSS	Mobility Services and Solutions
MTSO	Mobile telephone Switching Office
MTTR	Mean Time to Repair
MTU	Maximum Transitable Unit
MVNO	Mobile Virtual Network Operator
mVPN	Mobile Virtual Private Network
NAPTR	Naming Authority Pointer
NAS	Non-Access Stratum
NDC	National Dispatch Center
NENA	National Emergency Number Association
NF	Network Function
NFV	Network Function Virtualization
NG911	Next Generation 911
NIMS	National Incident Management System
NMC	Non-Mission Critical
NNI	Network to Network Interface
NOC	Network Operations Center
NOAA	National Oceanic and Atmospheric Administration
NPSBN	Nationwide Public Safety Broadband Network
NPSTC	Nationwide Public Safety Telecommunication Council
NS/EP	National Security Emergency Preparedness
NS/EP	National Security & Emergency Preparedness
NTIA	National Telecommunications and Information Administration
OA&M	Operations Administration and Maintenance
OAM	Operations Administration and Maintenance
OCS	Offline Charging System
OADM	Optical Add/Drop Multiplexes
OMA	Open Mobile Alliance
ONA	Optical Network Element
OPF	Originating Participating Functions
OS	Operating System

OSD	Operations Support Desk
OSPF	Open Shortest Path First
OSS	Operational Support Systems
OTA	Over the Air
OTT	Over the Top
PCEF	Policy and Charging Enforcement Function
PCI	Preemption Capability Indicator
PCRF	Policy and Charging Resource Function
P-CSCF	Proxy Call Session Control Function
PDCCH	Physical Downlink Control Channel
PDF	Policy Decision Function
PDN	Package Data Network
PE	Provider Edge
PGW	PDN Gateway
PII	Personally, Identifiable Information
PIV	Personal Identity Verification
PKI	Public Key Infrastructure
PL	Priority Level
PLMN	Public Land Mobile Network
PMIP	Proxy Mobile IP
PLRF	Proxy Location Routing Function
PNF	Physical Network Functions
PoC	Proof of Concept
POP	Points of Presence
PSAC	Public Safety Advisory Council
PSAP	Public Safety Answering Points
PSC	Platform Services Controller
PSE	Public Safety Entity
PSEN	Public Safety Enterprise Networks
PSTN	Public Switched Telephone Network
PTP	Point to Point
PTT	Push to Talk
PVC	Permanent Virtual Circuit
PVI	Preemption Vulnerability Indicator
QA	Quality Assurance
QCI	QoS Class Identifier
QoS	Quality of Service
QPP	Quality of Service Priority and Preemption
RAN	Radio Access Network

RAU	Routing Area Update
RCS	Rich Communications Services
RBDS	Radio Broadcast Data System
RF	Radio Frequency
RIM	Records and Information Management
RIR	Regional Internet Registries
RLP	Roaming Location Protocol
ROG	Response Operations Group
RRC	Radio Resource Control
RRH	Remote Radio Heads
RRU	Remote Radio Unit
RSRP	Reference Signal Receive Power
RSRQ	Reference Signal Received Quality
RTC	Regional Technology Center
RTP	Real-time Transport Protocol
RTT	Real Time Text
SABR	Service Access Border Router
SAE-GW	System Architecture Evolution Gateway
SAG	Service Aggregation Gateway
SAN	Storage Area Network
SBC	Session Border Controller
SCC-AS	Service Centralization and Continuity Application Server
SDK	Software Development Kits
SDLC	Software Development Life Cycle
SDN	Software Defined Network
SGi	Service Gateway Interface
SGW	Serving Gateway
SGSN	Serving GPRS Servicing Node
SIAD	Site IP Aggregation Device
SIAD	Smart Integrated Access Device
SIEM	Security Information and Event Management
SIM	Subscriber Entity Module
SIP	Session Initiation Protocol
SLA	Service Level Agreements
SLC	Satellite Link Case
SLO	Service Level Objective
SMS	Short Message Service
SNMP	Simple Network Management Protocol
SNOC	Security and Network Operations Center

SON	Self-Organizing Network
SPID	Subscriber Profile Identification
SPOF	Single Point of Failure
S/PGW	Serving/Packet Gateway
SPR	Subscriber Profile Repository
SRVCC	Single Radio Voice Call Continuity
TAI	Tracking Area Identifier
TAU	Tracking Area Update
TCA	Threshold Crossing Alert
TCP	Transmission Control Protocol
TDM	Time Division Multiplexing
TSP	Telecommunications Service Priority
TTY	Teletypewriter
UDP	User Datagram Protocol
UE	User Equipment
UI	User Interface
UL	Uplink
UMTS	Universal Mobile Telecommunications System
USIM	Universal Subscriber Identity Module
USP	Universal Service Platform
vCTS	Virtual Converged Telephony Server
vePDG	Virtual Enhanced Packet Data Gateway
VG	Virtual Gateway
VLAN	Virtual LAN
VM	Virtual Machines
VNF	Virtual Network Functions
VNS	Vehicular Network Systems
VoIP	Voice over IP
VoLTE	Voice Over LTE
VPLMN	Visited Public Land Mobile Network
VPN	Virtual Private Network
WEA	Wireless Emergency Alerts
WPS	Wireless Priority Service

References

1. Middle Class Tax Relief and Job Creation Act. [Online] 2012. www.congress.gov/112/plaws/publ96/PLAW-112publ96.pdf.
2. Subcommittee of the U.S. Congress. U.S. Government Publishing, ISBN 9781546636373. 2016.
3. U.S. Government. FirstNet and related Public Service Advisory Committee. [Online] www.firstnet.gov/about/psac.
4. U.S. Department of Interior. *RFP.* 2014.
5. 3GPP 23.401, Section 4.2. *General Packet Radio Service (GPRS) enhancements for Evolved Universal Terrestrial Radio Access Network (E-UTRAN) access.* 2015.
6. 3GPP. *Mobile World Congress, Barcelona, 19th February 2009.* 2009.
7. 3GPP TS 36.423, TS 36 01. *Evolved Universal Terrestrial Radio Access Network (E-UTRAN); Architecture description.* June 2010.
8. ITU-T E-212. *The international identification plan for public networks and subscriptions.* 2024.
9. 3GPP TS 36.311. *Radio Resource Control (RRC).* 2024.
10. 3GPP TR 21.905. *Vocabulary for 3GPP Specifications.* 2017.
11. 3GPP TS 23.401. *General Packet Radio Service (GPRS) enhancements for Evolved Universal Terrestrial Radio Access Network (E-UTRAN) access.* 2015.
12. 3GPP TS 23.402. *Architecture enhancements for non-3GPP accesses.* 2015.
13. 3GPP TS 23.203. *Policy and charging control architecture.* 2019.
14. 3GPP TS 24.229. *IP multimedia call control protocol based on Session Initiation Protocol (SIP) and Session Description Protocol (SDP); Stage 3.* 2019.
15. Internet Engineering Task Force (IETF). [Online] www.ietf.org/.
16. International Telecommunication Union (ITU). [Online] www.itu.int/en/about/Pages/default.aspx.
17. ETSI - TS 131 103 V6.6.0. [Online] www.etsi.org/deliver/etsi_ts/131100_131199/131103/06.06.00_60/ts_131103v060600p.pdf.
18. 3GPP TS 36.300. *Evolved Universal Terrestrial Radio Access (E-UTRA) and Evolved Universal Terrestrial Radio Access Network (E-UTRAN); Overall description; Stage 2.* 2017.

19. The Internet Society RFC 3888. *Message Tracking Model and Requirements.* 2004.
20. The Internet Society RFC 2960. *Stream Control Transimission Protocol.* 2000.
21. Internet Engineering Task Force RFC 761. *Transimmission Control Protocol.* 1980.
22. Internet Engineering Task Foce RFC 768. *User Datagram Protocol (UDP).* 1980.
23. IIETF RFC 768. [Online] 1980. www.rfc-editor.org/rfc/rfc768.
24. 3GPP TS 23.4 Section 5.3.2.1.
25. FirstNet. FirstNet Coverage. [Online] www.firstnet.com/coverage.html.
26. 3GPP TS 23.401. *General Packet Radio Service (GPRS) enhancements for Evolved Universal Terrestrial Radio Access Network (E-UTRAN) access.* 2015.
27. 3GPP TS 23.402. *Architecture enhancements for non-3GPP accesses.* 2015.
28. 3GPP TS 23.203. *Policy and charging control architecture.* 2019.
29. 3GPP TS 23.401 V18.6.0. 2024.
30. *6G Vision: An Ultra Flexible Perspective.* Yazar, A., et al. 1, 2020, ITU Journal on Future and Evolving Technologies, Vol. 1, pp. 1-20.
31. ITU-T. *G 872 – Architecture of Optical Transport Network.* 1998.
32. Singh, Madhu S. *Optical Transport Network (OTN): Operations Functional Requirements, SR 4616.* 1998.
33. ETSI TS 133.501. *Security architecture and procedures for 5G System.* 2018.
34. Metro Ethernet Forum. *Technical Specifications MEF 10.2.* 2009.
35. Internet Society. *Protocol Standards RFC 2474.* 1998.
36. Assured Forwarding Standards 2597, RFC. 1999.
37. 22.804, 3GPP SA1 TR. *Study on Communications for Automation in Vertical Domain R16 V16.3.0.* 2020.
38. The Internet Engineering Task Forc RFC 6665. *SIP-Specific Event Notification.* 2012.
39. 3GPP TS 24.229. *IP multimedia call control protocol based on Session Initiation Protocol (SIP) and Session Description Protocol (SDP); Stage 3.* 2015.
40. TS126.213, 3GPP. *Evolved Universal Terrestrial Radio Access (E-UTRA); Physical layer procedures.* 2022.

41. 3GPP TR 36.913. *Requirements for further advancements for Evolved Universal Rel 17.* 2022.
42. ETSI TR 25.913. *Requirements for Evolved UTRA (E-UTRA).* 2009.
43. (FCC), Federal Communication Commission. *Transition from TTY to Real-Time Text Technology.* 2016.
44. Register, Federal. *Transition From TTY to Real-Time Text Technology.* 2017.
45. 3GPP TS 22.226. *Global Text Telephony.* 2015.
46. 3GPP TR 23.204. *Support of Short Message Service (SMS) over Generic Internet Protocom (IP) access.* 2016.
47. 3GPP TS 23.057. *Mobile Execution Environment (MExE); Functional description.* 2011.
48. 3GPP TS 36.304 R 15.3.0. *Evolved Universal Terrestrial Radio Access (E-UTRA); User Equipment (UE).* 2019.
49. 23.229, 3GPP TS. *IP Multimedia Call Control Protocol based on SIP and SDP, Release 12.10.* 2015.
50. R12, 3GPP ETSI TS 29.212. *Policy and charging control over Gx reference point, Release 12.* 2015.
51. 3GPP TS 29.214 R12. *Policy and charging control over Rx reference point.* 2015.
52. 22.280, 3 GPPP TS. *Mission Critical Communications.* 2024.
53. 22.179, 3GPP TS. *Mission Critical Push to Talk (MCPTT); Stage 1.* 2024.
54. 22.281, 3GPP TS. *Mission Critical Video Services.* 2016.
55. 3GPP TS 22.282. *Mission Critical Data Services.* 2016.
56. R, 3GPP 23.246. *Multimedia Broadcast/Multicast Service (MBMS); Architecture and functional description.* 2015.
57. R14, 3GPP TS 23.280. *Common functional architecture to support mission critical services; Stage 2.* 2016.
58. R13, 3GPP. *Mission Critical Push to Talk Over LTE WID SP-150356.* 2015.
59. R14, 3GPP TS 23.379. *Functional architecture and information flows to support Mission Critical Push To Talk (MCPTT); Stage 2.* 2016.
60. 3GPP TS 23.167. *IP Multimedia Subsystem (IMS) Emergency Sessions.* 2015.
61. NENA-INF-016.2-2018. *National Emergency Number Association. Emergency Services IP Network Design for NG9-1-1.* 2018.
62. NINA-STA-031.1. *NENA Standard for Interconnecting Emergency Services IP Networks and Public Safety Broadband Networks.* 2021.

63. Code of Federal Regulation, Title 47- Part 10. *Wireless Emergency ALert.* 2024.
64. Federal Emergency Management Administration (FEMA). *Guide to Implementing the Integrated Public Alert and Warning System (IPAWS) V 2.0.* 2019.
65. (FCC), Federal Communications Commirssion. *FC 00-242.* 2000.
66. 3GPP TS 33.106 V18. *Lawful interception requirements.* 2024.
67. ETSI TS101.331 V1.4. *Requiremensts of Law Enforcement Agencies.* 2014.
68. European Telecommunications Standards Institute (ETSI) TS 102.232 V3. *Handover Specifications for IP Delivery.* 2021.
69. 3GPPP TS 23.501 V19.1.0. *System Architecture for 5G Systems (5GS) Stage 2.* 2024.
70. Ameigeiras, et al. *3GPPQoS-basedscheduling framework for LTE.* s.l. : EURASIPJournalonWireless Communicationsand Networking. DOI 10.1186/s13638-016-0565-9, 1916.
71. 3GPP TS 23.501 V19.1.0. *System architecture for the 5G System (5GS) Stage 2.* 2024.
72. V1`9, 3GPP 23.401. *General Packet Radio Service (GPRS) enhancements for Evolved Universal Terrestrial Radio Access Network.* 2024.
73. Blake S. et al. *RFC 2475. An Architecture for Differentiated Services.* 1998.
74. Rescue 42, Inc. *CRD for FirstNet – Users's Manual* [Online]. 2024.
75. IETF RFC 3263. *Session Initiation Protocol (SIP): Locating SIP Servers.* 2002.
76. Federal Eemergency Management Agency.
77. Rehmani, Mubashir H. *Blockchain systems and communication networks: From concepts to implementation.* May 2021.
78. Dylan Yaga et al. *Bockchain Technology Overview NISTIR 8202.* Gaithersburg, MD 20899-8930 : National Institute of Dtandards and technology - NISTIR 8202, 2018.
79. National Academies of Sciences, Engineering, and Medicine. *Emergency Medical Services: At the Crossroads.* Washington, DC: The National Academies Press. 2007.
80. CISCO. [Online] www.cisco.com/c/en/us/td/docs/net_mgmt/prime/network/3-8/reference/guide/sbcIMO.html.
81. 3GPP. [Online] www.3gpp.org/dynareport/23261.htm.
82. D.H.Morais. *5G and beyond wireless transport technologies.* s.l. : Springer; 1st ed. 2021 edition (July 9, 2022), 2022. p. 255. Vol. 1st edition.

83. NENA-STA-031.1. *NENA Standard for Interconnecting Emergency Services IP Networks and Public Safety Broadband Networks NENA-STA-031.1* . 2021.
84. V19.1, 3GPP 23.501. *System architecture for the 5G System (5GS);*. 2024.
85. E-212, ITU-T. *The international identification plan for public networks and subscriptions.* 2024.
86. 3GPP TS 24.229. *IP multimedia call control protocol based on Session Initiation Protocol (SIP) and Session Description Protocol (SDP); Stage 3.* 2015.
87. 24.264, ETSI TS. *IMS-based AR Real-Time Communication.* 2022.
88. 24.229, 3GPP TS. *IP multimedia call control protocol based on Session Initiation Protocol (SIP) and Session Description Protocol (SDP).* 2019.
89. 23.401, 3GPP. *[R1] 3GPP TS 23.401, Technical Specification Group Services and System Aspects; General Packet Radio Service (GPRS) enhancements for Evolved Universal Terrestrial Radio Access Network (E-UTRAN) access, Release 10.0.* 2010.
90. Internet Engineering Task Foce RFC 768. *User Datagram Protocol.* 1980.
91. Rescue 42. *CRD handbook.* 2020.
92. ITU-T G.107. *E Model.* 2020.
93. Engineering and Operations in the Bell System, Bell Laboratory, Library of Congress Catalog No. 77-84418, 1977.

Index

Note: Figures are indicated by *italics*. Tables are indicated by **bold**.